Soul-Searching

Soul-Searching

The Evolution of Judeo-Christian Thinking
on the Soul and the Afterlife

Luke Jeffrey Janssen

FOREWORD BY
Malcolm Jeeves

WIPF & STOCK · Eugene, Oregon

SOUL-SEARCHING
The Evolution of Judeo-Christian Thinking on the Soul and the Afterlife

Copyright © 2019 Luke Jeffrey Janssen. All rights reserved. Except for brief quotations in critical publications or reviews, no part of this book may be reproduced in any manner without prior written permission from the publisher. Write: Permissions, Wipf and Stock Publishers, 199 W. 8th Ave., Suite 3, Eugene, OR 97401.

Wipf & Stock
An Imprint of Wipf and Stock Publishers
199 W. 8th Ave., Suite 3
Eugene, OR 97401

www.wipfandstock.com

PAPERBACK ISBN: 978-1-5326-7981-0
HARDCOVER ISBN: 978-1-5326-7982-7
EBOOK ISBN: 978-1-5326-7983-4

Illustrations used with permission.

Cover: "The Last Judgement," a bas-relief residing in Autun Cathedral, France, believed to have been created around 1130 AD. In this scene, the Devil and the Archangel Michael weigh human souls (lined up along the lower right) to determine who will go to heaven (up the left side of the scene) or to hell (down the right hand side). I felt this image was perfectly suited to draw the reader's attention to my offering on this subject. The figures and images themselves overtly speak the key themes of the book: the soul and the afterlife. As well, the sense of antiquity and the sacred with which this image is imbued resonate with the historical and theological nature of the book's contents. The observer will likely find that they need to take the image in slowly and parse it carefully in order to fully comprehend it: the same will be true of what they will find inside this book.

Manufactured in the U.S.A.   08/02/19

Contents

Foreword by Malcolm Jeeves | vii
Preface | xi

Chapter 1: Introduction | 1
Chapter 2: Premodern Human Ontology: The Mind, Soul, and Spirit | 31
Chapter 3: A Synthesis of Modern Thinking on Our Inner Being | 71
Chapter 4: The Afterlife | 129
Chapter 5: Change | 184

About the Author | 203
Bibliography | 205
Subject Index | 219
Scripture Index | 231
Ancient Document Index | 241

Foreword

Malcolm Jeeves

CBE, FMedSci, FRSE

Emeritus Professor, School of Psychology and Neuroscience, St. Andrews University

Past President of the Royal Society of Edinburgh, Scotland's National Academy

The past half-century has witnessed remarkable advances in our understanding of the brain. The last decade of the twentieth century, labeled "The Decade of the Brain" by the US Senate, resulted in increased funding for brain research of all kinds. By the turn of the twenty-first century, researchers realized the possibility of actually seeing which areas of the brain are most active when volunteers were doing all sorts of tasks, such as looking at art, listening to music, showing maternal love, meditating, or praying. Everything seemed well set for yet further rapid advances in the study of mind and brain, leading some scientists to suggest that the first decade of the present century should be called "The Decade of the Mind and Brain." Very soon, with the widespread use of smart phones and similar devices, it became customary to talk about the software and the hardware of such devices. This way of thinking about the relationship between mind (the software) and brain (the hardware) seemed to make good sense. It served further to underline the unity of the device being used and by implication the unity of the human person. It made sense to see mind and brain as two essential aspects of the one unity: the human person.

But what about the soul? Is the soul the same as the mind? If not, how does it differ? For two millennia a pervasive theme of dogmatic and systematic theology when focusing on theological anthropology and the doctrine of humanity emphasized that humankind alone is created "in the divine image" or "in the image of God." This refers, of course, to the book of Genesis chapter 1 verse 27 where we read, "God created man in his own image, in the image of God he created him." On this view, it was held that a straightforward answer to the question of what makes us human and distinguishes us from the rest of creation was that, since God is a spiritual being, He endowed us also with spirituality, giving us an immortal soul. That, however, turns out to be too simple. Any reference to the writings of biblical scholars and theologians who have traced out the understanding of the concept of soul over more than two millennia demonstrates the wide variety of views that have been taken over that period, and, as Janssen documents, indeed before.

Few contributors to the ongoing debate about the nature of the soul and the human person have the knowledge and background from both within science and within theology to make a sustained attempt at working out a view of the human person, and in particular of the soul, which does full justice to advances in scholarship in both theology and science. The author of this present book is one such person. His distinguished career as a medical scientist with a deep understanding of the almost unbelievable intricacy of our biological makeup, and, at the same time someone who has made the effort to undertake further specialized training in biblical scholarship and theology, equips Professor Janssen to guide us through the minefields of competing theories about our human nature and specifically about the soul.

Of the many excellent features of this volume, two are particularly noteworthy since they illustrate how the author has made reference to and underlined the range of relevant evidence, whether within the domain of biblical studies and theology or of contemporary science, which speak to a deeper understanding of our mysterious human nature including the soul. First, he underlines the need to recognize that *Homo sapiens* left Africa approximately 70,000 years ago via what is sometimes called the Levantine corridor—a narrow stretch of land adjoining Africa to Eurasia—and landed first in what is now the Near East before expanding to the rest of the world. Views of the soul did not begin with Hebrew and Greek views. They were preceded by the influence of the Babylonians, the Assyrians and the Persians, each of which left a mark and reshaped not only Judaism but also the development of Christianity and hence views of the soul. Second, from a position of firsthand knowledge as a medical researcher over several decades, he reminds the reader that "our bodies have in fact been in a continuous process of dematerialization and rematerialization throughout our whole lives!" and that "the unique body which I have today is not, to be precise, the same one I had ten years ago, and will not be the same one I possess in another ten years." Nevertheless, he concludes that "philosophers and scientists are rejoining theologians in the belief that we are more than just the material. There is a seemingly immaterial component—consciousness, both individual as well as universal—which we just cannot yet explain."

Strong claims and counterclaims have been made about the merits of so-called monism and dualism as models of the human person. Janssen offers us a detailed and evenhanded account of the evidence in support of monism and of dualism. On balance, he comes down on the side of monism but throughout, at no point, understates the strength of the dualist case. A view I share and which has led my view to be characterized as "dual-aspect monism." For him, as for many of us still working on this topic, the important message is "keep an open mind and seek to do full justice to advances both in the relevant sciences and in the fields of biblical scholarship and theology." This is not a book for those who want slick and simplistic answers to profound and difficult questions. It is a book that will guide the thoughtful reader to a deepening understanding of our mysterious human nature.

Preface

I was raised in a Fundamentalist Christian setting which placed a high priority on the words of the Bible—an ancient document—as interpreted through the lens of a sixteenth-century Western worldview (given to us by the Reformation) that underwent a few relatively minor adjustments during the five centuries that followed its appearance on the world stage. This setting and worldview worked just fine for me during the first three decades of my life. However, going to university in pursuit of a career in science, while also being exposed to other entirely new (to me) philosophical and religious ideas, changed everything for me. After descending into an ever-deepening state of agnosticism which occasionally dipped into atheism, I began to look for a Christian faith that might be consistent with modern science and a view of world history that is not biased by certain theological presuppositions. I have detailed that deconstruction and ongoing reconstruction in a self-published book entitled *Reaching into Plato's Cave: Bringing the Bible into the 21st Century*. When I refer to an "unbiased view of history," I mean one which accepts at face value the overwhelming evidence of humans arising out of Africa and migrating across the planet over the course of countless millennia, rather than starting with *a certain interpretation* of Genesis set in the Mesopotamian basin less than ten thousand years ago and then trying to fit around that a carefully selected subtotal of whatever scientific data I could find to be compatible. That alternative paradigm is the focus of a second book—entitled *Standing on the Shoulders of Giants: Genesis and Human Origins*—which summarizes the scientific data pointing to human descent from other hominid ancestors, and then reexamines Christian theology in light of that.

One of the lessons I learned in researching both books was that we humans have been on a journey stretching out over hundreds of thousands of years to find the Divine. That search had us always looking at the eastern horizon, toward the rising of the sun which seems to have always represented new beginnings and supreme power, and which may have been one of the impulses that drew us out of Africa and into the new lands which stretched out before us. One of the major routes we took was via a narrow stretch of land connecting Africa to Europe: Egypt on the African side, and that which we now call the Middle East on the Eurasian side. Those migrants—our genetic and theological/philosophical ancestors—rested their feet in Mesopotamia for a while as they planned the next stages of our collective journey into Europe and Asia (and then eventually the Americas and Australia). As they did, they continued to look outwardly to understand the divine, as well as inwardly to understand the essential core of our inner being: the human soul. That geographical region—the ancestral lands of Babylon, Egypt, Persia . . . as well as Israel—is precisely where we now find the world's oldest remains of temples and religious texts: it is the birthplace of human writing, and an incubator of religious thinking.

Coming to this realization opened up for me a whole new perspective on many of the biblical passages which had long been so puzzling or even upsetting for me. This spurred me to dig more deeply into the development of our thinking about the human soul. The central core of this book is a summary of what I learned: how that thinking transformed over the many millennia; how an ancient Near Eastern worldview that dominated the entire region for millennia may have influenced the emerging theology of a man from Babylon named Abram, and his many descendants who eventually gave us Judaism; how that in turn was transformed during the Babylonian captivity and by classical Greek philosophy, eventually becoming Christianized; and finally how Judeo-Christian ideas were challenged severely in their encounters with modern science and the emergence of more critical thinking. All of this is summarized in chapter 2 of this book, and sets the stage for chapter 3 in which I explore modern thinking about the human soul, including the science underlying consciousness, the mind, personality, soul, and spirit (five very different aspects of the human experience).

But much of the content of those two chapters may be foreign, perhaps even threatening, to my primary intended audience: Christians with relatively little training in world history, philosophy, and biblical theology. They may base much of their worldview on the words on the pages of their Bibles, but without an awareness of the contexts of the writing of those biblical texts, nor an appreciation of the roles played by other humans in the writing *and interpretation* of those texts. Most readers will know that it was not lowered down fully formed out of heaven; but just the same, many will have little or no idea of how we *did* receive it. They might have vaguely formed mental images of authors sitting at desks, writing specific words that they heard audibly or somehow perceived subconsciously. An informed understanding of this process and journey of spiritual discovery is crucial for a more complete picture of what those words were intended to mean to the original hearers/readers of the biblical narratives, and also their meaning to us in the modern context. So before presenting chapters 2 and 3, I felt the need to set the stage for the reader: to describe how we came to receive the Bible as we now have it. Thus: chapter 1.

Having in this way laid out a full history of the Judeo-Christian understanding of the (immaterial) soul which is intimately tied to our very material bodies, it made sense to explore what happens when we encounter that one inevitability in life: death (I disagree with Benjamin Franklin that "taxes" are the other experience of which all humans can be certain). It turns out that the Judeo-Christian understanding of the afterlife has also been constantly transforming over the past many millennia. Some might say it has been maturing, or opening like a flower. At any rate, chapter 4 summarizes this part of our theological history as well. Finally, chapter 5 ties many of the ideas together and raises a number of questions regarding how this applies to us in the twenty-first century.

Let me emphasize, the goal of this book is *not* to give the final authoritative definition of the soul or the afterlife. I leave that to the readers to formulate their own answers for themselves, only now with some additional background material on the table. Instead, I hope to show how our understanding of both has changed tremendously over the past several thousand years, and seemingly in response to the thinking of culture(s) all around us. A secondary goal will be to raise awareness of other viewpoints on these two questions, and to encourage dialogue without confrontation. Related to this, I want to show that a modern understanding of the human mind and soul does not have to be fundamentally different for a Christian compared to a nonbeliever (reader please note: this statement does *not* apply to our understanding of the human spirit, a completely different matter): in fact, no more different than their respective views on the nature and origin of the *physical* aspect of our existence. We may have our differences on certain key theological points, but it seems that some Christians too often act as if we must disagree with non-believers on *everything*, and become alarmed when it is claimed that we can find common ground. When it comes to biological origins, Christians and nonbelievers alike can and should find it possible to explain the appearance of humans using similar natural mechanisms (gradual biological evolution, not an instantaneous shaping of clay). As my Old Testament professor, Dr. Gus Konkel, asked me: "Is not evolution simply a somewhat sophisticated way of describing the sculpting of clay or mud?" Likewise, Christians and nonbelievers can just as easily use natural mechanisms to explain the origin of the cosmos (big bang cosmology), or how babies are formed (not via knitting needles, as per Psalm 139), or the weather and climate (not the way these are described in Job chapters 37 and 38). And now we can have similar naturalistic explanations for the human soul. All of these statements are testimony to the fact that resorting to natural, physical explanations does *not* need to jeopardize Christian faith. In fact, I find that reconciling scientific discoveries with a biblical theology presents a much *greater* God than simply believing he acts only through mystical, magical means. The Book of God's Word and the Book of God's Works indeed!

A tremendous amount of the thinking that went into this book was done in the context of taking courses at the McMaster Divinity College, leading to the completion of a master of theological studies degree. In fact, many sections which now appear in this book were first written as essays for some of the courses I took while studying there. I am immensely grateful for the instruction, mentorship, camaraderie, intellectual stimulation, and coffee chats enjoyed with the professors and students there. Special thanks in this respect go to my fellow student Ambrose Thomson and to Drs. Paul Bates, Mark Boda, Paul Evans, Gus Konkel, and Christopher Land. I want to reiterate an acknowledgment I made in my previous book toward all the students and faculty at the McMaster Divinity College—coming from such diverse backgrounds and having so many different faith journeys—for opening up a whole new perspective on Christian belief, for putting up with (and even sometimes responding to) my many outrageous questions in class, and for helping me find a Christian faith which I can now easily hold up side-by-side with my secular vocation in medical research and general interest in all things scientific.

I also benefitted greatly from innumerable discussions with fellow members of the American Scientific Affiliation and its Canadian partner the Canadian Scientific and Christian Affiliation. Presentations made at the annual ASA meetings and the monthly CSCA meetings, as well as a 2017 meeting hosted by the BioLogos Foundation in Houston, Texas, have unfailingly been a source of new ideas, new collegial relationships, and inspiration.

I am greatly indebted to several other individuals for being sounding boards and discussion partners as I explored this subject and bounced ideas off their well-read and critically-thinking minds: Scott Dyer, Paul Almas, Rev. Steve Baldry, and my siblings, Allan Janssen and Margot Casuccio. Scott, Steve, and Allan in particular were an immense help in reading drafts of this monograph and providing critical feedback from the point of view of a nonexpert (the intended target audience for this book). Likewise, Dr. Konkel read the final draft and provided feedback as an expert scholar of the

Old Testament and ancient Near East, while Dr. Ken Post played the same role as a scholar of theology and philosophy.

Many of the ideas in this book were also tested in a blog post forum that I maintain at https://lukejjanssen.wordpress.com/. A large number of people dialogued with me there—some friendly, others not so much—and gave me much food for thought. They are too many to name, and perhaps several would not wish to be named. But I thank them just the same.

Finally, and most importantly, I want to thank my wife, Miriam, for indulging my new hobby and vocation, listening to my thoughts, reading my writings, and invariably providing new perspectives and insights into what I'm saying or how I say it. It was in fact she who proposed the title for this book, after I had spent weeks casting around with a number of lesser alternatives. She is indeed the iron which sharpens me.

CHAPTER I

Introduction

In ancient Egyptian mythology dating back to the middle of the third millennium BCE, the soul of the dead would be weighed in a balance against the feather of Ma'at: the person's eternal destiny would be determined by that weighing. Well over a millennium later, Job would plead, "Let God weigh me in honest scales and he will know that I am blameless" (Job 31:6). A millennium or two later yet, Roman Catholicism developed a tradition of Michael the Archangel weighing human souls in a balance on the judgment day (see cover). Fast-forward yet another millennium, we find a twentieth-century physician in Massachusetts placing dying patients on weigh scales in an attempt to determine the mass of their soul at the moment of death. We will revisit all four stories in later chapters of this book, but they are juxtaposed here to highlight a recurring theme: our fascination with defining the human soul.

Human origins (anthropogony) and human existence (ontology) have been subjects of particular interest across all societies in all periods of history. This has been especially true for theists. In the fourth century, St. Augustine prayed: "What then am I, my God? What is my nature?"[1] Millennia before him, the psalmist wrote: "What is man that you are mindful of him?" (Ps 8:4). Long predating both, the mythical hero Gilgamesh intrigued ancient Babylonians with his story of embarking upon a quest to discover the secret to immortality and the afterlife. Today, through our intellectual and scientific sophistication, we continue to cultivate an interest in this subject of human existence. Hollywood has long fueled this discussion with movies that explore aspects of human origins, of the core essence of what it means to be human, and of immortality (*2001: A Space Odyssey*; *Contact*; *Prometheus*; *Ex Machina*; *Transcendence*; *Blade Runner*; *Her*; *The Discovery*; *I, Robot*). Needless to say, innumerable books have been written on this subject. Is there a ghost in the machine, or are we just molecules

---

1. Augustine, *Confessions*, XI.

bouncing around? This is very much an important question that literally strikes at the heart of our existence.

My primary goal in this book will not be to give a final definition, description, or explanation about our inner being or the afterlife, but rather to show how Christianity arrived at the position(s) that we now find ourselves to hold: to trace a journey which has been unfolding over thousands of years. And in the process of traversing this path, how we constantly borrowed ideas and theories from others around us, and were influenced by our cultural, philosophical, and even religious environments. Christians may claim to derive their understanding on these questions from the Bible. Any readers who pause over the words "claim to derive" here might ask themselves how much time they actually invest in *studying* the Bible to explore these questions *for themselves*, versus how much of their understanding is merely parroted from hearing what others have to say about the matter (their religious leaders and/or fellow believers, who in turn are parroting what they have heard from yet others, and so on). Rather than beginning a sentence with "The Bible says . . . ," they might want to consider substituting that with "My pastor says . . . ," or "My parents taught me . . . ," recognizing that the latter in turn have likely simply heard the same from yet others before them.

Furthermore, irrespective of the extent to which readers consult the Bible for their understanding, many believers do not fully understand the origins of Scripture itself, nor appreciate the extent to which various Christian traditions have derived widely different interpretations of those writings.

For these reasons, it will be illuminating to explore briefly in this opening chapter three key points: (1) the cultural influences which constantly acted on the unfolding and evolution of Judeo-Christian thinking; (2) the authorship of religious texts; and (3) the process by which the Bible came to be. For some, coming to understand those three points in detail for the first time can be quite disorienting; it may lead them to have doubts about Scripture. Those who come to that inner conflict should learn to distinguish between what is actually written in the Bible and what they were led to *believe* is written in the Bible (by well-meaning pastors, parents, teachers, peers, and others), and should also seek to learn how certain concepts were understood very differently in a cultural context completely different from our own. It is often said that those of us in twenty-first-century Western society would do well to remember that the Bible was written *for* us, but not *to* us.[2]

---

2. Walton, *Lost World of Genesis One*, 7.

Cultural Context in Which Judeo-Christian Thinking Took Shape

Many Christian and Jewish scholars accept that Abraham—the ancestral patriarch of the nation of Israel, and the spiritual father of Judaism, Islam, and Christianity—lived sometime around the beginning of the second millennium BCE. (Estimates for dates such as this one vary markedly among scholars: being more specific does not advance my argument any further and only creates a red herring to dispute over. I hope the reader will indulge imprecise statements such as this one in which relative timeline comparisons are being made.) Those scholars would further agree that the bulk of the texts upon which Judaism and Christianity are founded were written by Abraham's descendants during the next fifteen hundred years, a period which includes Israel leaving captivity in Egypt to become an independent nation (approximately the middle of the second millennium BCE), establishing her own monarchy and temple-centered religion (approximately the turn of the first millennium BCE), having her political and religious systems being utterly destroyed by the neo-Babylonian empire (approximately the middle of the first millennium BCE), and the reestablishment of a new national identity and religious system during a rapid succession of global superpower empires. We will see below how those empires (neo-Babylonian, Assyrian, Persian, Greek, and Roman) and several later ideological forces left their mark on the reshaping of Judaism and the development of Christianity. However, even the earliest Hebrew writers, and Abraham himself, did not live in an ideological vacuum. Long before them, the Near East was already dominated politically, culturally, and intellectually by several other civilizations, and it was saturated with their religious ideas. With this thumbnail sketch of the history of the ancient Near East, we will now look more closely at the political and religious incubator which produced Judeo-Christian thinking.

It is important to recognize that *Homo sapiens* left Africa approximately seventy thousand years ago via the Levantine corridor—the narrow stretch of land which joins Africa to Eurasia—and landed in what is now called the Near East before expanding throughout the rest of the world. We now have overwhelming evidence of our ancient African origin and those human migrations (including other launching points further down the eastern coast of Africa), which will not be presented here; many other sources are available for that information,[3] and I have reviewed it extensively in a previous

---

3. Ayala, "Human Nature," 31–48; Tattersall, "Acquisition of Human Uniqueness," 25–42.

book which also explores the impact of this new information on Christian theology.[4] This makes the relatively small patch of land that we now call the Near East or the Middle East one of the oldest incubators of modern human philosophical and theological thinking. Over the course of dozens of millennia, as humans entered and passed through that narrow stretch of land between Africa and the rest of the world, they began to generate various mythologies, cosmogonies (theories describing the origin of the universe), anthropogonies (theories describing the origin of humans), and ontologies (theories about existence, and what it means to be human), borrowing ideas from others that they encountered in their wanderings and social interactions and passing the net product on to subsequent generations.

Eventually, certain people groups founded large cities in the Mesopotamian basin, which grew to dominate the entire region around them, including the people of Sumer (Sumerian civilization) in the late fifth millennium BCE, Egypt in the fourth millennium BCE, and the people of Akkad (Akkadian civilization) in the third millennium BCE.[5] Many biblical scholars accept that the first eleven chapters of Genesis bear a striking resemblance to mythological ideas held by those surrounding ancient Near Eastern civilizations for millennia before those Hebrew chapters were written. Those civilizations (and Genesis) share in common the belief that the cosmos began as a primordial watery mass out of which the land was separated, and that humans were created from some form of dirt.[6] At the twelfth chapter of Genesis and onward, however, many scholars see actual Jewish history, beginning with the introduction of Israel's patriarch, a Babylonian named Abram. Even if this is true, the surrounding superpowers of that ancient Near Eastern world had been developing their mythologies, theologies, and philosophies for several millennia before Abram appeared on the scene.

The Torah tells us that Abram was thoroughly Mesopotamian in his origin and upbringing: he came from Ur of the Chaldeans (Gen 11:31), and his family originally "worshiped other gods" (Josh 24:2) and had a corresponding Mesopotamian mythology. Abram's family would have been very familiar with that mythology: for example, we have evidence that, at a later period, the Babylonian mythological *Epic of Creation* (*Enuma Elish*) was recited, and possibly enacted, on the fourth day of every New Year festival.[7] Abram took his immediate family on a long journey from Ur to Canaan. At one point

---

4. Janssen, *Standing on the Shoulders*.

5. Quirke, *Exploring Religion*, 206.

6. Watts, "Making Sense," 5–7; Longman, *Proverbs*, 205; Whybray, *Proverbs*, 128–29; Waltke, *Book of Proverbs*, 415.

7. Dalley, *Myths*, 231; Levenson, *Creation and the Persistence of Evil*, 69.

along the way, one of Abram's clan is said to have taken her family idols with her (Gen 31:19, 34); it seems that others also kept their gods, since much later in their story, Jacob "said to his household and to all who were with him, 'Get rid of the foreign gods you have with you'" (Gen 35:2). Although those gods were eventually discarded, it is impossible to ascertain to what extent the Mesopotamian zeitgeist was passed on through the successive generations of children raised by the mothers who sang to their babies the lullabies they themselves learned when they were young: we know of no proscriptions against them doing so, and we have no reason to think that they suddenly stopped singing those lullabies or repeating those stories. We are never told that YHWH revealed to Abram/Abraham an entirely new cosmogony and anthropogony distinct from that of the Babylonians.

In fact, we also do not know when or how YHWH might have revealed a wholly different theology to Abram, and it is possible that Abram and his descendants were still working that out while journeying to the promised land of Canaan. For example, Abraham (his name had now been changed; Gen 17:5) told Abimelech that "*elohim*" caused him to wander from his homeland (Gen 20:13). "*Elohim*" is a plural noun which is often used in the Old Testament (well over two thousand times). When scholars find this plural noun followed by a verb in the singular form, they routinely translate it as "God"; when it is followed by a verb in the plural form, it is taken to refer to the gods of other nations (Exod 12:12; 1 Kgs 11:33), to angels (Ps 8:5), and to the spirit brought forth to King Saul by the medium at Endor (1 Sam 28:13). Except in this singular occurrence of the conversation between Abraham and Abimelech. Abraham referred to "*elohim*" causing him to wander, and the Hebrew text has him using a *plural* form of the verb: but breaking away from their rigorous practice of otherwise translating this as "*the gods* caused me to wander," scholars instead interpreted Abraham as saying "*God* caused me to wander" in order to avoid a theological dilemma. This exchange with Abimelech took place *after* Genesis describes "*elohim*" appearing to and speaking with Abram several times over the course of twenty-five years: when Abram was first called out of his homeland (Gen 12:1); after dividing the land with Lot (Gen 13:14–17); when God reiterates his covenant with Abram (Gen 15:1); when God gives Abram/Abraham the covenant of circumcision and foretells that the promises would come through Isaac, not Ishmael (Gen 17:1–22); and when Abraham met three visitors at his tent the day before the destruction of Sodom and Gomorrah (Gen 18). One might think that, after all these theophanic encounters, Abraham would have formed some kind of understanding of "*elohim*," except for the fact that "*elohim*" had appeared to him in different forms, including a smoking firepot with a blazing torch (Gen 15:17) and a physical human who shared a meal with Abraham at the latter's tent (Gen 18:1–8), let alone other possibilities which are not explicitly stated (A disembodied voice? A light? An angelic presence?). So why would Abraham have used this plural verb form with Abimelech? One suggestion might be that he was simply accommodating the polytheistic worldview of Abimelech. That may be the case: but it is equally possible that Abraham was born and raised as a Babylonian, and was still working out his new and entirely different theology. Again, we have no records of YHWH revealing a detailed theology which was so different from everything that Abram had been taught during the first three-quarters of a century of pondering the divine, and was now recounting to Abimelech a couple decades later.

Any views this family had regarding anthropogony and ontology were probably orally transmitted (we have no indication that they carried any texts with them on their nomadic journey through the desert) and now seem to be lost to history: the Old Testament speaks only vaguely on this subject. However, oral tradition would always precede any texts themselves, and although it is impossible to date an oral tradition, we know that the origins of the surrounding Babylonian culture predated the period in which Abraham lived by several *thousand* years. It would not be surprising, then, that Abraham's nomadic family members were also heavily influenced by the Mesopotamian ideas that they grew up with and about which they were reminded whenever they encountered others along their journey. Many such encounters are given in the book of Genesis, including extended ones with Canaanites (Gen 12:6; 23:1–2; 34–35), Egyptians (Gen 12; 39–50), Amorites (Gen 14), Sodomites (Gen 14; 18), Hittites (Gen 23), Philistines (Gen 20–21; 26), and Arameans (Gen 33). Although the book of Genesis does indeed describe several direct encounters between YHWH and the patriarchal heads of this nomadic family, those encounters are brief and not accompanied by lengthy or detailed theological discourses; once again, as I wrote above for Abram/Abraham, one can only conjecture whether YHWH revealed to them any details of any cosmogony, anthropogony, or theology which were entirely new and different from those of the surrounding nations.

A few generations later (early to middle second millennium BCE), this nomadic family is said to have settled in the heart of the Egyptian empire (Gen 47:27; Exod 1:1–7), which had its own religion and mythology, and its own distinct view of anthropogony. The Egyptians believed in an afterlife, at least for their pharaohs and select (usually wealthy) individuals, but not in the concept of resurrection for all people.[8] According to Old Testament texts, Abraham's descendants remained there in Egypt for centuries—biblical

---

8. Charlesworth, "Origin and Development of Resurrection Beliefs," 219.

authors give this as 400 years (Gen 15:13; Acts 7:6), 430 years (Exod 12:40; Gal 3:17), or about 450 years (Acts 13:20)—and were eventually led out of Egypt by Moses to settle in the land of Canaan. We have no record, biblical or otherwise, to tell us that at this time they had a view of human ontology and final destiny which was different from that of the Egyptians.

Prior to the most recent and in-depth scholarly research, it had been thought that the first five books of the Hebrew and Christian Bibles—the Pentateuch—were written largely by Moses in the second millennium BCE.[9] Assuming for the moment that that was indeed the case, what do we know of that author? Moses is said to have received an extensive Egyptian education (Acts 7:22). As a member of Pharaoh's household (Exod 2:10; Acts 7:21; Heb 11:24), he would have been taught by Egyptian scholars, who would have instilled in him *Egyptian* ideas regarding human origins and the afterlife. During that schooling, he would undoubtedly have accessed the Egyptian libraries, which included ancient Sumerian and Akkadian texts. All of that exposure to ancient Near Eastern ideology would certainly have influenced any textual writing that might be attributed to Moses.

And what could one say about any text that he is claimed to have written? It is worth pointing out first that the evidence available to us at present informs us that the system of Hebrew writing itself did not begin to take shape until approximately the beginning of the first millennium BCE.[10] So whatever texts would have been in Moses's possession centuries before that would have been in some Mesopotamian language. One might suggest that the language he used was probably some form of Aramaic, the *lingua franca* of that place and time which eventually gave us Hebrew, Arabic, Phoenician, and several other scripts. But we cannot state that definitively, since we have not had the original texts for millennia. Could it be that Moses wrote his own texts, at least some of them, in an Egyptian script?

In contrast to this view of Mosaic authorship, we will consider a contemporary scholarly view that the Pentateuch was markedly revised, and many other books of the Old Testament actually written, over a millennium later during and after the neo-Babylonian exile. But first we need to consider Hebrew theological development during the intervening years between the time of Moses and the time of the neo-Babylonian exile (well over a thousand years of development: consider how much our own philosophy and theology changed between the year 1000 CE and the present). The books of Deuteronomy, Joshua, Judges, Chronicles, and Kings summarize

---

9. Waltke, *Genesis*, 22–28.

10. Cahill, *Gifts of the Jews*, 6; Sanders, *Invention of Hebrew*, 103; Finkelstein and Silberman, *Bible Unearthed*.

that period of Hebrew history in which the nation of Israel extended herself out across Canaan. Some modern scholars add that various indigenous Canaanite, Edomite, and Transjordanian people groups amalgamated with the Hebrew immigrants from Egypt and together formed the nation of Israel,[11] but that is a matter of dispute outside the scope of this monograph. Those Old Testament texts describe various military leaders (Joshua, Gideon, Deborah, Jephthah, and others), religious prophets (Samuel, Eli, Elijah), and otherwise influential individuals (Samson) leading them through crises and events. Those stories may or may not be historical, but they do explore theological questions and set precedents on which later theological development would be founded.

It is easy for us today to read those stories and imagine a homogenous nation of Israel acting collectively as one unified consciousness, moving immediately from one event or crisis to another. This is an artifact produced by our modern reading of the texts, moving instantly from one event or story to the next as we turn the pages of our Bibles. However, we need to keep in mind that any two stories may have taken place in completely different and remote parts of Israel, with large geographical separations and gaps of time in between them, let alone large cultural divides. That is, in a time before radio, television, and internet, the theological proclamations of a Samuel or a Nathan might not have been heard by Israelites living in villages even just one hundred miles away; instead, those distant villages might possibly be influenced more by the proclamations of local prophets representing entirely different ideologies. For readers who might not fully accept my claim here, consider this: even in today's context which *does* have those electronic tools which homogenize society (such as radio, television, newspapers, and the internet), a small village in a remote rural area of any modern country might have its own set of superstitions, legends, "common-sense," patterns of speech, and even ideology relative to the more central and densely populated parts of that country, and even more so relative to another small village on the other side of that country (for example, consider two remote villages in Ireland and Scotland, or in Alabama and Tennessee). This geographical and temporal separation in ancient Israel—which would foster ideological diversification—would explain (in part) why we perceive in the Old Testament texts an image of the pre-monarchical nation of Israel vacillating wildly in her worship of YHWH, Baal, Asherah, or Molech; or descending into civil war (Judg 12:1–6; Judg 20:1–48); or bouncing around from one ideology to the next.[12] The book of Judges twice highlights this earlier lack of

---

11. Toorn, *Family Religion*, 181.
12. Toorn, *Family Religion*, 183–205.

unity and cohesion: "In those days Israel had no king; everyone did as they saw fit" (Judg 17:6; 21:25).

As van der Toorn put it: "The emergence of a state religion under King Saul inaugurated a new phase in the cultural history of early Israel. Until then, religion had been diverse: each family or clan had its own religion, honored its own god, and celebrated its own rites."[13] The establishment of a monarchy and national army, and the building of the religious temple, however, all contributed to a uniting of the Israelite people, which resulted in a greater homogenization of Hebrew ideology and theology.[14] Texts written at this time might better reflect what the nation as a whole was thinking (although there could still be significant differences between what the nation as a whole claimed to represent and what individuals actually thought and practiced).[15]

Following the descent of the Israelite monarchy into political and religious anarchy, the Jews were taken captive into Babylon by King Nebuchadnezzar early in the sixth century BCE. Whatever state or national religion which existed in Israel prior to that event was now ended. Modern scholarship holds to the view that the Pentateuch and many of the Hebrew texts were written and/or radically edited at this time, in part in an attempt to make sense of the dramatic changes that were taking place. What were the philosophical, religious/theological, and political forces acting on Jewish thinking at that time?

The Jewish nation felt it had lost favor with her God, and they had broken the covenant between them: "What was YHWH now going to do?" they were asking. This was followed by deep national and personal sorrow and introspection, and considerable prophetic activity and theological anguish (as per the prophets Isaiah, Jeremiah, Ezekiel, and many others). Several decades later, the socio-political-religious landscape suddenly changed. Cyrus the Great, who is referred to by Ezra (1:1) and the author of 2 Chronicles (36:22), conquered the neo-Babylonian Empire and returned the Jews to their ancestral lands, allowing them to rebuild their religion. From the Jewish perspective, YHWH had in this way given Israel a second chance. And so they quickly set themselves to the task of rebuilding their destroyed religious system, including extensive writing/copying of religious texts. This newly revised version of the religion that they produced is generally referred to by modern scholars as "Second Temple Judaism" (because the first temple had been destroyed).

---

13. Toorn, *Family Religion*, 287.
14. Toorn, *Family Religion*, 181–82.
15. Toorn, *Family Religion*, 1–4.

Quite contrary to what some might otherwise think, there is extensive evidence that the Hebrew scribes did not simply copy and reproduce verbatim the ancient religious texts, including the Pentateuch, which is often attributed to Moses. Readers who think that way probably envision copyists working with an original manuscript on one side of their desk, a blank scroll or page on the other, meticulously transferring the text letter-by-letter from one side to the other while one or more other scribes looked over their shoulder or at the very least performed careful comparisons and tests to ensure that the reproduced text was a "carbon-copy" of the original. Scenarios like that one may have been true in a much later era (for example, the rabbinic scribes rebuilding and copying the Masoretic texts in the seventh to tenth centuries CE), but a large body of evidence tells us this was not at all the case during the exile. Many modern biblical scholars have described the extensive editing and rewriting of Jewish texts which took place during this period in Israel's history, as they tried to make sense of the dramatic turn of events which had occurred since the covenantal promises made to Abraham, Moses, and David. The revisions included: invention of entirely novel passages with ideas that were not present in the original texts; expansion of existing passages with new embellishments; compilation, integration, and adaptation of other passages, sometimes even from texts which are now not considered to be Scripture.

Scholar of ancient religions Karel van der Toorn describes in detail one example of this. He gives evidence of the book of Deuteronomy being essentially a series of books within books, concatenated like Russian dolls. The oldest of these was an ancient text—the Torah of Moses or the "*Urdeuteronomium*"—purportedly written by Moses and then long lost but rediscovered in the seventh century BCE by Hilkiah the priest, who was scouring the temple for gold and silver. This text then underwent four successive revisions every few decades. Each revision involved extensive insertions of new text written by various groups of scribes, including an additional new prologue and colophon (opening and closing statements, respectively) for each revision, leaving elements of the preceding prologues and colophons in place as "vestigial remnants" (which then later serve as markers of this historical development for scholars who know what to look for). Van der Toorn refers to these successive versions of Deuteronomy as the "Covenant Edition," "Torah Edition," "History Edition," and "Wisdom Edition," respectively.

Van der Toorn also provides a sense of the political motives behind these revisions. The very original text pertained to a time when Israel was her own nation, and was ruled by a royal and/or military leader who was the supreme ruler in the land of Israel. Even when the *Urdeuteronomium* was rediscovered in the seventh century, Israel was ruled by King Josiah just as the nation was descending into full-blown captivity by the Babylonian and Persian empires. By the end of this series of editorial revisions, however, the final revision of Deuteronomy (the one which we now possess) describes the priests as the highest court of the land: even the king is required to receive a copy of the Torah from their hands and to read from it daily. Power had shifted from the monarchy to the priesthood!

A further political dimension is added to the process when one considers that the development of this text was triggered by directives from the Persian Empire. Part of their *modus operandi* in ruling their conquered nations was to have the vanquished write up a code of law based on their own traditions and religious practices. Provided that the tribal codes of law did not conflict with Persian law, it was those newly written/revised texts which became the law of the citizenry of that defeated nation. This gave those conquered people a sense of autonomy, as well as a motivation to not rise up in rebellion lest they risk losing the relative peace and stability they had (Cyrus was a smart man!). But from a Judeo-Christian point of view, this also meant that the new Hebrew laws and texts which were written had to first pass through a Persian filter. Near the end of his book, van der Toorn writes: "Without the Persians, there would not have been a Pentateuch. The Persians are responsible, too, for the transformation of the Torah into the Law."[16]

The Jewish scribes of that postexilic era who did that writing and text-editing would have had access to the libraries of the Babylonian and Persian empires (which would also have contained copies of the ancient Sumerian, Akkadian, and Egyptian mythological texts referred to above), and frequently would have been exposed to the oral accounts and reenactments of mythology from the perspectives of those other civilizations in addition to that of the Persians.

More importantly, in addition to exposing the Hebrew religious leaders—who were busy rediscovering and redeveloping Judaism—to the texts and ideas of those other ancient Near Eastern civilizations, their close encounter with Cyrus and the Persian empire put them in direct contact with Zoroastrianism,[17] which originated in Persia possibly as early as 1500 BCE. One of several characteristics which distinguished Zoroastrianism from the other ancient Near Eastern religions was a very well developed belief in the afterlife. The Old Testament describes the prophet Daniel being inducted into the political-religious center of this Persian Empire by Cyrus himself

16. Toorn, *Scribal Culture*, 251; also see Toorn, *Family Religion*, 352–62; note: Jonker calls for a careful and nuanced approach to assessing "the Persian context" in biblical studies.

17. Martin-Achard, *Death to Life*, 186–93.

(having already been educated and trained by the intelligentsia of the previous Babylonian Empire; Dan 1:1–6, 19): he experienced their religion, politics, and zeitgeist directly and intimately. We will see in chapter 4 that Daniel suddenly introduced to the conquered Jews a new perspective on the afterlife, and that his writings are the first in the Hebrew Bible to explicitly introduce the concept of resurrection (some claim his book is *the only* one in the Hebrew Bible to do so). Likewise, Nehemiah is central to the rebuilding of Jerusalem and Second Temple Judaism, and his story begins with him serving as cupbearer to Artaxerxes, king of the Persian Empire (Neh 1:11), a high-ranking position which would put him in the inner circles of power in that pagan empire. Zoroastrianism and Judaism are similar in many respects (especially with respect to the origin and end of the world, the afterlife, and spiritual beings [angels; Satan]), and they share the distinction of claiming to be the first monotheistic religions.

In the late fourth century BCE, the Persian (Achaemenid) Empire fell to the Greeks under Alexander the Great, and Judaism succumbed to an intense Hellenization.[18] Serious readers should avail themselves of the extracanonical Hellenistic Jewish literature which appeared during this intertestamental period, since it will provide a new perspective on Jewish thinking around the time of Christ. These other books may not be considered Scripture to us today, but those texts were indeed read and discussed by the Jews of Jesus' era, including Paul and the disciples. Greek civilization arose from those of the Minoans and the Mycaenians in the early to middle second millennium BCE; their ancient mythology is similar in many respects to that of the ancient Babylonians and Egyptians. However, Greek society had changed dramatically by the middle of the first millennium BCE: during the Hellenic and Hellenistic periods, they moved away from mythicism and toward much more advanced schools of philosophy.[19] They believed in a supreme transcendent being or force (*Logos*) that did not involve itself with earthly existence, and in immortal souls which are trapped in material bodies. Some of the world's greatest philosophers—Socrates, Plato, Aristotle, Pythagoras—transformed the worldview of this Greek global superpower, and then the Roman superpower that followed on its heels. Classical Greek philosophy continues to influence philosophical and theological thinking right up to today.

In the first century CE, Christ introduced a completely new interpretation of Hebrew Scripture—"you have heard it said . . . but now I say . . ."—and the New Testament church expanded on that dramatically. The Apostle

---

18. Green, "Restoring the Human Person," 8–11; Green, "Bodies," 161.
19. Burkert, *Greek Religion*, 305–37.

Paul, with his extensive Greco-Roman Hellenistic training, is particularly responsible for that expansion of Christianity's new theology. With the fall of Jerusalem and the destruction of her temple late in the first century CE, Second Temple Judaism diverged in two very different directions—rabbinic Judaism and Christianity—both of which continued to be transformed by classical Greek philosophy.

At this point, the reader might be chafing in their chair with the constant interjection, "But what about divine inspiration?" and thinking I have downplayed or even ignored the role of the Holy Spirit in the writing of the biblical texts. Yes, indeed, the Holy Spirit was inspiring the writers and thinkers. But I think the church too easily downplays the influence that the surrounding culture exerted on them as well; in fact, it may not recognize that the Holy Spirit might be using those other cultures as a lens to reveal truth (we will look at this idea more closely on pages 22 and 190). This external influence explains the dogmatic prohibition imposed by the New Testament church against eating meat offered to idols (Acts 15:29) and then their coming to reconsider and reverse that prohibition (1 Cor 8:4-13), or Paul's dogmatic statements about hair length for men and women (1 Cor 11:14-15) which we now completely dismiss. It is a matter of debate—and the focus of this book—where one draws the line between divine inspiration and human discovery or invention.

It is worth pointing out that many (most?) of the earliest Christians read the Bible in Greek, and interpreted it in a *Hellenized* Hebrew culture, if not a thoroughly Greco-Roman one. Over the next few centuries, various "fathers of the church" (such as Ignatius, Polycarp, Irenaeus, Clement, Augustine, to name just a few from a very long list) saturated the nascent Christian theology with Platonic/Neoplatonic ideas, and a sprinkling of some Aristotelian ones. We will see later in the next three chapters how these catalyzed a profound reinterpretation of ancient Hebrew Scripture and drove early Christian thinking into completely new directions.

In summary, then, according to the Old Testament texts themselves, the nation of Israel traces its founding father from within a Mesopotamian zeitgeist (Gen 35) that influenced global thinking for millennia, who led his small clan of wandering desert nomads to a new global superpower with its Egyptian zeitgeist (Gen 46). Many centuries later, that clan, now grown into a nation, is led by a thoroughly Egyptian-trained leader to Canaan (Exod 12) where they develop their own unique religion while immersed in a variety of other Canaanite cultures. Only much later yet do they develop their own unique language to fully articulate their story and their theology. A millennium later, that nation finds itself conquered and scattered and its religion crushed, but then gathered together again within a neo-Babylonian zeitgeist, then a Zoroastrian/Persian worldview, and finally a Hellenistic Greek perspective in which they *literally and completely* rebuild their religious system and its accompanying religious texts.

Undoubtedly, the latter were shaped by the zeitgeists of those earlier global superpowers. By their sheer size, wealth, military strength, and advanced technologies, those other civilizations influenced all the nations within their reach, including worldviews and philosophical/religious ideas. All texts are written in a cultural language: that is certainly true today, and no less true thousands of years ago. As a Canadian, I can relate directly and personally to the power of that kind of cultural influence, seeing on a daily basis the impact that American media, culture, economics, history, and zeitgeist have on my fellow Canadians and our society. We Canadians may sometimes "borrow" outright certain ideas and values from them, but in a far greater number of ways we are almost always responding to their ideas, values, and narratives.

Parallels between Hebrew Scripture and Ancient Near Eastern Writings

Some readers may resist the possibility that the Old Testament and Christian authors were influenced in their writing by their surrounding cultures, and possibly even "borrowed from" the latter at times. But we can easily find evidence of that influence nonetheless: this is most easily done by comparing stories found in the first eleven chapters of Genesis (although there are *many* examples in other parts of the Bible) with legends held by ancient civilizations that preceded the Hebrews. Here are five examples.

The Creator God

For the ancient Sumerians—several thousand years before the era of Abram and Moses—Enlil was the most prominent of the many gods birthed by the sky-god An and the earth-goddess Ki.[20] Enlil split apart heaven and earth, and ruled over all the other gods from within the space he so created. His name derives from the Sumerian word for "breeze," "wind," and "breath," and was thought to be the force that allows living beings to exist. We will see in the next section below that the Sumerians and Egyptians believed that a collaboration of gods mixed clay with a god-derived life force to create humans. Could these legendary ideas have been reinterpreted by later Hebrew authors

---

20. Watts, "Making Sense," 5–6.

describing God's wind-like spirit hovering over the waters before separating the waters above from the waters below, as well as the land from the water, and then uttering that puzzling statement using the plural form—"Let *us* create mankind . . ." (Gen 1:26; italics added for emphasis)—before breathing the breath of life into a lump of clay to create the first man?

Likewise, the ancient Egyptian texts refer to: the god Ptah speaking things into existence and separating earth and sky; the god Re creating light out of darkness; a firmament separating the upper and lower waters; various gods creating animals, fish and birds; and creating humans from clay (see next example).[21]

The Creation of Man from Clay

Sumerian and Akkadian mythology describes humans being created from clay, including: *Song of the Hoe*; *Hymn to E'engura*; *Enki and Ninmah*; *KAR4*; *Enuma Elish*; and *Atrahasis*.[22] In *Atrahasis*, composed in the twentieth to sixteenth centuries BCE (although oral versions would have long preceded this), the mother goddess Nintu and the god Enki mix the blood and flesh of a slain god (*Geshtu-e*) with clay, giving humans a dualistic nature, comprising the earthly and the divine.[23] In *Song of the Hoe*, and in the *Hymn to E'engura*, seeds are planted and humans break out from the ground like crops.

The ancient Egyptians also asserted that humans were created out of clay by their own deities, although this seems to be more of an afterthought or an accident than an intentional act on the part of the gods.[24] In one late third millennium BCE myth, tears shed during a sorrowful encounter between Ra-Atum and his two daughters turn into human beings when they hit the dirt floor, while in another equally old myth, humans are molded out of clay on a potter's wheel by the god Khnum.[25] In both stories, humans are only peripheral details, and no myth describes any kind of direct or intimate relationship between humans and the gods.

Many ancient Greek texts describe the creation of all things, but the most detailed is given by the poet Hesiod within *Theogony*.[26] They saw a long

---

21. Watts, "Making Sense," 6–7.
22. Noort, "Creation of Man and Woman," 13–4, 17–18; Janssen, *Standing on the Shoulders*, 43; Dalley, *Myths*, 4, 15–6, 228–77; Watts, "Making Sense," 5–8.
23. Abusch, "Ghost and God," 364.
24. Watts, "Making Sense," 6–7.
25. Noort, "Creation of Man and Woman," 13–14; Quirke, *Exploring Religion*, 55, 143; Pedersen, *Israel*, 99.
26. Bremmer, "Pandora," 20–29.

recurring series of gods engaging in conflict and creating other lesser gods and then various mortal beings; in each case, they endowed their creations with different abilities and powers.[27] In the last of these creation events, the god Prometheus made man out of mud, and the goddess Athena breathed the spirit of life into the man (again, a collaboration of gods). Since all the good skills and qualities had already been given away to the previous creations (such as swiftness, cunning, strength, fur, wings), nothing was left to give the human. So Prometheus gave humans the ability to stand upright like the other gods (in other words, "in their image"), and gave them the ability to create and control fire. Another Greek myth had Deucalion (son of Prometheus) and Pyrrha (daughter of Pandora) repopulating a flood-ravaged earth by throwing stones over their shoulders.[28]

In all four cases—in the Sumerian, Akkadian, Egyptian, and ancient Greek civilizations, which spanned several millennia of human history, and which exerted political-cultural-sociological influence in their respective eras—we see a common theme: a collaboration of gods creating humans by mixing dirt with some kind of divine ingredient (blood, flesh, tears, breath). Is it surprising, then, when we read in Genesis—written by authors who lived within that ancient Near Eastern context—that "the Lord God formed a man from the dust of the ground and breathed into his nostrils the breath of life, and the man became a living being" (Gen 2:7)? Job makes the same claim for himself (Job 10:8–9), and the psalmist also seems to echo it: the familiar passage in Ps 139:14 ("I am fearfully and wonderfully made") is followed by the often overlooked or omitted phrase "when I was woven together in *the depths of the earth*" (Ps 139:15; emphasis added). This puzzling wording in Psalms becomes particularly relevant when we learn that the ancient Mesopotamians saw the underworld (the depths of the earth) as the abode not only for those who had died but also where the unborn child (the *kūbu*) lived until it transitioned to the mother's womb,[29] or that the Egyptians believed that all of creation emerged from the deep darkness of the earth.[30]

Adam is explicitly said to have been formed "from the dust of the ground" (Gen 2:7). Part of the curse that was put on him for having broken God's command to not eat from the Tree of the Knowledge of Good and Evil was a reminder of that fact: "for dust you are and to dust you will return" (Gen 3:19). Walton claims that this reference to the man being fashioned from dust was not intended to be taken literally: that this was not describing

---

27. Larson, *Understanding Greek Religion*, 68.
28. Montgomery, *Rocks Don't Lie*, 158.
29. Scurlock, "Mortal and Immortal Souls," 79.
30. Watts, "Making Sense," 6.

material origins of the human body, in part because "the Israelites would not be inclined to thinking in terms of chemistry."[31] It is true that "chemistry" is a concept and a scientific discipline which would not exercise human thinking for yet many more millennia. However, the Hebrews, like many other ancient societies, *would* nonetheless make a material association between dust and the body (and death). Many ancient societies would leave their dead in open spaces that were not easily disturbed. The Egyptians and their pyramids are a widely recognized example of this, but we will see later in chapter 4 how the ancient Israelites constructed bench-tombs in which the body was laid out as if in sleep. In both cases, the body would decay in their exceptionally dry climate, leaving only bones and dried flesh. The ancient Hebrews would return long after the fact in order to gather up the bones into jars or ossuaries which held the bones of other ancestors: in the process of stirring up the dried remains, the flesh would crumble into dust. If there had been any kind of drafts in those tombs, some of the dried remains would have already been wafted up and settled on various surfaces of the tomb. They would recognize the bones and dust as the residual components of the body, just as they would connect rubble and heaps of bricks with the walls of derelict buildings, and ashes in a firepit as the remains of a burnt tree trunk. So in their minds, the body and death were intimately tied with dust. Even Abraham, the patriarch of all the Jews, said: "Now that I have been so bold as to speak to the Lord, *though I am nothing but dust and ashes*" (Gen 18:27; emphasis added). It is not surprising then, that they would also envision an omnipotent Creator God reversing the chaotic forces of death, refashioning that dust back into a body and infusing it with a vital force or living spirit.

The Association between Woman and the Fall of Mankind

At least one Mesopotamian myth parallels closely the story of the fall described in the third chapter of Genesis. The biblical story begins with Adam and Eve living a blissful life in the garden, in communion with the animals. They directly break a Divine command in an attempt to gain wisdom (the knowledge of good and evil). Although they were originally created in God's image and likeness, God now remarks to the Divine Council that "the man has now become like one of us," at which point they are clothed with the skins of animals and ejected from the garden.

The Mesopotamian version, on the other hand, begins with a wildman (Enkidu) created by the gods and living a solitary but Eden-like existence in

---

31. Walton, *Lost World of Adam*, 72.

communion with the animals. Enkidu is transformed by a sexual encounter with a harlot (Shamhat), as a result of which he loses his fellowship with the animals: the latter flee because he has lost his animal-like "innocence" and has gained reason, knowledge, and wisdom. Now fully human, he puts on clothes and has become like a god.

One ancient Greek myth describes Zeus wanting to punish the god Prometheus for having done wrong (deceiving Zeus), so he collaborated with Hephaestus to create his own destructive deception: the first woman. Pandora was stunningly beautiful, and Prometheus's own creation—the first man—was irresistibly drawn to her (as Zeus and Hephaestus fully expected). But she had been created with a deceptive heart and a lying tongue, and was given a jar which she was commanded not to open. When temptation overwhelmed her and that command was disobeyed (again, as her creators fully expected), she unleashed a series of misfortunes, diseases, and plagues upon humans. There are two distinct and intriguing theological parallels between this ancient Greek myth and Genesis chapter 3. First, we see the first woman as the ultimate cause of the destructive forces that mar human existence, by allowing curiosity and selfish desire to move her to break a divine command, and then inviting her husband to do the same: this echoes the passage from which certain Christians have derived the concept of original sin (Judaism never did make this leap). Second, the follow-up to this Greek myth contains elements of salvation from an eternal punishment that we later see in Hebrew and Christian theology. Prometheus was also punished: bound up with unbreakable chains and tormented day and night by an eagle that each day ate at his liver. Each night the liver regenerated, allowing this torment to continue the next day. The eternal punishment could be reversed if an immortal would volunteer to die for Prometheus and if a mortal killed the eagle and unbound him (Chiron the Centaur and Heracles later filled these roles). In the biblical story, Adam's seed (a mortal) would trample the head of the snake in the garden, and Christ (also Adam's seed, and an immortal) volunteered to die in the place of the condemned.

Cain and Abel

One Sumerian myth parallels the biblical one of Cain and Abel. Enten came into conflict with his brother Emesh when they brought thank offerings to the supreme god Enli. Enten, who attended to irrigation of crops, brought agricultural produce (animals, plants, wine, and beer), while Emesh brought a very different offering (gold, silver, lapis lazuli, wood, and fish). Although the Sumerian story does not end in fratricide, the brother who lived closer to the land was favored by Enlil. This resonates with the story in Genesis of God favoring Abel's animal offering for reasons which are never explained in the biblical texts.

Noah and the Global Flood

There are many close and undeniable parallels in the Genesis account of the Great Flood and several ancient Near Eastern flood narratives which preceded it by millennia.[32] In particular, the Sumerians developed the story of Ziusudra in the early third millennium BCE, which was adapted by the Akkadians a millennium later with the key figure in the story renamed Atrahasis, and again a few centuries later, the story was adapted by the Babylonians within the *Epic of Gilgamesh*, now featuring the key figure named Ut-napishtim.[33] In this story, the gods become bothered that the humans they had created were becoming too numerous and were making too much noise at night with all their working, so they decided to drown them in a flood. A very Noah-like character was instructed by a sympathetic and protective god to build a boat and load it with his family and animals to ride out that flood. After the storm had passed, that survivor of the global flood released three different kinds of birds before leaving his boat and offering a sacrifice to the protective gods, who in turn were well pleased with the smoke of the sacrifice. (If the reader is not aware of how closely these narrative details parallel the Genesis account, they would do well to review the biblical flood account.)

The ancient Greeks also had their own version of such a story, in which Zeus attempted to destroy humans by drowning them in a flood. The sole survivors of that flood—Deucalion (son of Prometheus) and Pyrrha (daughter of Pandora)—repopulated the devastated earth by throwing stones over their shoulders. However, another ancient Greek myth features Zeus feeling threatened by the humans he had just created and wanting to destroy them.[34] In this story, primeval humans are described as being round like a ball and having two faces, four arms and hands, and four feet, which enabled them to be always moving, always doing, and always seeing. More importantly, they were immortal and highly proliferative, growing to incredible numbers. Zeus again began to feel threatened and wanted to destroy them, but reasoned that they were too valuable for the worship that

32. Heidel, *Gilgamesh Epic*; Damrosch, *Buried Book*; George, *Epic of Gilgamesh*; Dalley, *Myths*, 50–153; Watts, "Making Sense," 5; Montgomery, *Rocks Don't Lie*, 143–60.

33. Waltke, *Genesis*, 132.

34. Plato, *Symposium*.

they offered to the gods and the work that they did. His ingenious solution was to cut these spherical humans in half, which resulted in less powerful subjects, but twice as many worshippers.

All three stories—the Mesopotamian, Greek, and Hebrew—feature the divine becoming concerned about humans increasing in numbers, and deciding to solve the problem by drowning them in a flood. In two of these stories, another divine solution to the human problem was to cut back human vigor (whether that be literally cutting their bodies in half [in the Greek version], or shortening their existence [in the Hebrew version]). Given that the Mesopotamian version was written millennia before the Hebrew narrative, and that the ancient Greek version might also predate the writing of Genesis, it would be reasonable to suggest that the Hebrew author(s) were aware of one or both of those other versions, and possibly influenced by them.

As an aside, the author of Genesis introduced that narrative with a "problem": the human population "began to increase in number on the earth" (Gen 6:1) and was consorting with "the sons of God" (Gen 6:2). The Divine solution to the problem(s) was to limit human existence to a hundred and twenty years (Gen 6:3), and I have often heard or read it pointed out how human lifespans plummeted after the flood event. Could this puzzling introduction to the flood narrative be simply a way for the Hebrew authors to reconcile their story with Mesopotamian and Greek myths of such interactions between gods and humans and a global flood? Rather than drowning humans immediately as the Sumerian/Akkadian gods did, or literally cutting the transgressing humans in half as Zeus did, YHWH would drastically cut back on the length of their existence. Below (page 45), we will consider a different interpretation of this puzzling passage.

Inspiration or Plagiarism? Or Both?

In all five cases, we find similarities in certain details of the stories and plotlines, but also in the underlying values, perspectives, assumptions, and presuppositions. They spoke (and wrote) in the language of their culture. We will see below how all these ancient societies associated life and soul with one's breath (which betrayed their inner emotions and which departed from them, sometimes audibly, upon death). They would likewise experience the wind that swept across the face of the earth and associate that with the Divine breath, and even with the Divine per se. They would see dead bodies slowly rot and sink into the ground if left in the open, or turn into dust if left in a cave or an artificial enclosed space, and thereby associate the body with dirt, mud or clay, and the Creator(s) shaping that dirt and infusing it with their life-giving wind. Their society was inherently paternalistic, and took a certain view of "the woman." Like most premodern societies, their cities needed to be built near a reliable source of drinkable water. That often meant living beside a massive river like the Nile or the Euphrates (which they had deified) which annually swelled over their banks; every now and then the overflow took on massive proportions with correspondingly overwhelming destruction, and this would remind them of a time when the flood was seemingly "global" and must have been a consequence of the deities punishing them.

The Jewish authors, and their hearers/readers, were immersed in this ancient Near Eastern zeitgeist all around them. They might have thought that way as well; or if not, they would have felt a need to respond to it. It is easy to see, then, how the parallels between these pagan myths and passages in Genesis might be examples of Jewish authors actively responding to and adapting the mythology of the cultures around them, either at the time of their first writing and/or when those texts were being revised during the exilic and postexilic periods.

Some believers today maneuver around this by suggesting that the Hebrew authors did indeed have those pagan traditions and mythologies in mind when they wrote their own texts, but did so as a polemic against the former rather than as simple plagiarism (polemic involves the exercise of arguing against someone else's ideas).[35] For example, while there are clear similarities in the creation accounts of the Hebrews and the other civilizations, there are important differences. The gods of the other cultures were powerful, but not all-powerful: they could lose control of the flood waters, or of the chaos monster, were subservient to the *kittum*, or unchangeable forces of the cosmos, and could be killed. YHWH, on the other hand, was in a *completely* different league. Those other gods all emerged from matter of some kind and were dependent upon it, while YHWH *created* all matter. Genesis describes everything being created as an act of love and relationship-building rather than one of cosmic/divine warfare (Akkadian; Sumerian) or a sorrowful accident (Egyptian). Genesis also describes a very intimate relationship between God and humans, rather than one of servitude or even slavery to the gods. In the Hebrew version, *all* humans bear the image of God, not just the king (as was the case for the pagan civilizations). The Mesopotamian gods want/need food to be brought to them: YHWH provides for the humans a garden full of food to enjoy. The things which

---

35. Clouser, "Reading Genesis," 237–61; Pinnock, "Climbing Out," 143–55; Harlow, "Creation according to Genesis," 163–98; Fugle, *Laying Down Arms*, 241; Sanders, *Invention of Hebrew*, 76; Watts, "Making Sense," 5–12.

other civilizations viewed as gods to be worshipped and feared (mountains, seas, sun, moon, stars, animals) are the very things which YHWH creates for human service and enjoyment.

There certainly can be an element of polemics here. But polemic is not just about writing words down: it is about confronting a conversational adversary. Like other literary forms—propaganda; love songs; hate speech; hagiography—polemic only works if it actively seeks, encounters, confronts, and challenges another viewpoint. All of these literary forms lose their *raison d'être* if they are kept private. To my knowledge, Genesis was never distributed to the Babylonian or Egyptian scholars of their time; neither did the Hebrews in that era engage in debates between opposing views of the kind that we see the church fathers doing during the Patristic period (nor as internet-warriors do in our own era). They did not fight against the curriculum being taught in their children's schools. The authors/editors of Genesis and their contemporary readers/listeners were quite intellectually, sociologically, and religiously cloistered.

That having been said, there may be room for a much softened version of the word "polemic." That is, while they may not have been *actively opposing* the contemporary worldviews (true polemic), they may have been *simply responding to and adapting* the latter (closer to syncretism): the Hebrew writers might have been also formulating their own religious thoughts and interpreting through the lens of ancient Near Eastern ideology, consciously or subconsciously. This can even be seen as a form of divine inspiration: God placed these ideas and philosophies in front of the ancient Hebrews as a lens by which to gain an entirely new perspective on their emerging theology. This suggestion need not be perceived as a threat to the Christian believer: there is an important difference between "influence" and overt "plagiarism." We can see other examples of "influence" in biblical writings. All of the biblical texts are attributed to male authors, and their writings are very clearly male-oriented and male-centered. All of the ancient authors were prescientific, and modern scholars note that nothing in Scripture advanced scientific understanding beyond the science of the world in which they were written.[36] The Hebrew writers were also influenced by their own human emotions: perhaps the most disturbing example of this would be Psalm 137:9 ("Happy is the one who seizes your infants and dashes their heads against the rocks"). Ancient Hebrew historians wrote in the same way that others did in that era: for example, it made perfect sense to them to introduce a story as occurring "in the spring, at the time when kings go off to war" (2 Sam 11:1; see also 1 Chr 20:1), or to manipulate and telescope a genealogy in order

---

36. Walton, *Lost World of Genesis*, 19.

to reinforce a particular theological point. It makes sense that the Apostle Paul, with his first-century Jewish perspective, would have such strong views about the disgrace of a man with long hair, or of a woman with short hair (1 Cor 11:1–16). So it should not be surprising that the writings that address their philosophical/theological ideology would also be *influenced* by conversations and ways of thinking that were going on all around them, and by stories which the authors grew up with. Their writings reflect their ancient Near Eastern context (and later, a Hellenistic Greek one).

On the other hand, might it be that God was also working through those other pagan civilizations: in their own pagan search for the Divine, they caught glimpses of truth which they incorporated into their own mythology and which the Hebrew authors noticed, recognized, latched onto, and together with their own revealed glimpses of truth developed further under the inspiration of the Holy Spirit? Early church fathers such as Justin Martyr, Irenaeus, Origen, and Clement of Alexandria indeed taught that God was revealing aspects of his truth to the pagan nations.[37] Justin Martyr wrote of the *logos spermatikos*, the seeds of truth that God had put in all people: this concept was instrumental as a bridge between Greek philosophy and emerging Christian thought. Clement went so far as to say that Christianity was a divinely orchestrated amalgamation or marriage of Judaism and Greek thinking:

> God made a new covenant with us; the ones with Greeks and with Jews are ancient. But we worship him in a new way as the third race. For clearly, as I think, he showed that the one and only God was known by the Greeks in a pagan way, by the Jews in a Jewish way, but by us in a new and spiritual way.[38]

I have often found it striking how human civilizations throughout all of history and across every part of the globe have been seeking out the divine. Religious impulses are felt by people of *every* race, gender, age, political leaning, level of wealth, and sociological status. I believe that yearning to find and understand the Divine is itself Divinely inspired. So humans across the globe searched, and contemplated, and shared their ideas with each other: in the process, they developed ever more elaborate religious structures. Some of them were sensitive in varying degrees to the promptings of the Holy Spirit. We would naturally want to include the Hebrew and Christian authors in that list of those being so prompted, but I do not see why we cannot also accept that God was working through pagans as well, in

---

37. McDermott, *God's Rivals*, 85–156.
38. Ashwin-Siejowski, *Clement of Alexandria*, 8.

much the same way that Scripture itself describes God acting through the pre-Hebrew priestly figure Melchizedek (Gen 14:18; Ps 110:4; Heb 5:6, 10; 6:20; 7:1–2, 10–11); the Phoenician widow in Zarephath (1 Kgs 17:9; Luke 4:26); Naaman the Assyrian commander (2 Kgs 5:1–19; Luke 4:27); Cyrus the Persian emperor (2 Chr 36:22; Ezra 1:1–8; Isa 44:28); the Magi (who came from "the East"; some suggest these were Zoroastrians) celebrating the birth of "the Hebrew King" (Matt 2:1–2); or the Ethiopian eunuch (Acts 8:26–39); in fact, Scripture claims God could even work through Balaam's donkey (Num 22:21–30). During his visit to Athens, Paul identified the God whom the pagan Greeks were seeking to worship, even quoting truths from their Stoic philosophers, and said, "God did this so that they would seek him and perhaps reach out for him and find him" (my paraphrase of Acts 17:22–30). Augustine thought similarly:

> Pagan learning is not entirely made up of false teachings and superstitions. It contains also some excellent teachings, well suited to be used by truth, and excellent moral values. Indeed, some truths are even found among them which relate to the worship of the one God. Now these are, so to speak, their gold and their silver, which they did not invent themselves, but which they dug out of the mines of the providence of God, which are scattered throughout the world, yet which are improperly and unlawfully prostituted to the worship of demons. The Christian, therefore, can separate these truths from their unfortunate associations, take them away, and put them to their proper use for the proclamation of the Gospel.[39]

In his exploration of how Christ's suffering was a means of our deliverance, N. T. Wright writes "that such an idea—one person suffering to redeem many—was widespread in the ancient non-Jewish world, turning up in Homer, Euripides, and many other famous non-Jewish writers as well as in speeches from heroes in battle."[40] Other scholars have also written about non-Jews in the Old Testament and non-Christians in the New Testament perceiving truths about God.[41] Below, we will see clear evidence of the development of Hebrew theological ideas in response to non-Hebrew philosophies: Daniel suddenly introducing a whole new concept of resurrection during/after his time spent in Zoroastrian Persia, or the fathers of the church radically developing Christian theology after being thoroughly trained in classical Greek philosophy.

---

39. Augustine, *De doctrina Christiana*, II.xl.60–61.
40. Wright, *Day the Revolution Began*, 125.
41. McDermott, *God's Rivals*, 27–42.

Many contemporary Christian scholars and theologians will point to hints and vague elements of a given theological concept in much older texts, sometimes even in texts from non-Hebrew sources. And in doing so, those contemporary writers will at times employ biological wording related to germination, fertilization, or fetal development to emphasize that those somewhat subliminal hints later developed into a full theology.[42] I have already mentioned above Justin Martyr's reference to the *logos spermatikos*. Likewise, while researching the concept of resurrection, I found the following quotations (I have added italics for emphasis):

- "In the earliest texts in the Old Testament we can find *pregnant indications* of the later development of the belief in the return of the departed to life."[43]

- "*Two seeds* lay in *fertile soil* awaiting the *moistening dew* that would act as a catalyst to promote *germination*"[44] (the two seeds being theological concepts, and the catalyst being persecution, injustice, martyrdom, theodicy).

- "Each of the several elements that appear in Daniel 12:1–3 existed, *at least germinally*, in earlier stages of the religion of Israel, though their combination and fusion are relatively new and distinctive."[45]

- The theological basis for a resurrection theology can be found latent "*in embryonic form.*"[46]

- "The Iranian doctrine helped the Jews to *adopt a conception* whose essential elements had already been provided by their own tradition, and perhaps to widen it, or even to transfigure it, so that it might give the answer to a question which, for generations and especially after the Exile, had haunted the minds of the Israelites."[47]

Some readers might prefer to limit any *human* influence on the authorship of the Hebrew texts, attributing it instead largely (some, perhaps even entirely) to YHWH. However, that viewpoint can raise interpretational difficulties. One good example of the latter arises from a consideration of the commonalities between the various ancient Near Eastern texts and Genesis regarding "the woman" who contributes to the demise of "the man":

42. McGrath, *Science of God*, 227.
43. Martin-Achard, *Death to Life*, 188.
44. Crenshaw, "Love Is Stronger," 55–56.
45. Madigan and Levenson, *Resurrection*, 199.
46. Routledge, "Death and the Afterlife," 22.
47. Martin-Achard, *Death to Life*, 193.

- In the book of Genesis, Eve is deceived to take from the tree of the knowledge of good and evil and then invites Adam to do the same in order to gain wisdom, followed by: (*i*) the loss of a peaceful existence in paradise in communion with the animals; (*ii*), the introduction of sin and death; (*iii*) God needing to kill an animal(s) to clothe them;
- In Enkidu's encounter with the prostitute Shamhat, Enkidu gains wisdom and knowledge and becomes like the gods, but in the process loses fellowship with the animals because he kills one to clothe himself;
- In the ancient Greek myth, the first woman Pandora is provided to man by the gods as a punishment or even a tool-of-destruction, and unleashes all manner of evils and calamity upon the first man.

In this case, then, forcing the thesis that the Sumerians, Akkadians, ancient Greeks, and Hebrews were all divinely inspired when writing about "woman" and "evil" raises difficult questions. It might be easier to attribute this commonality to the fact that the authors in all four societies were predominantly (likely exclusively) male with an ancient Semitic attitude about women.

On the other hand, to those who would suggest that the borrowing went in the other direction—that the commonalities between the pagan mythologies and the Hebrew texts is a result of the pagan cultures in fact borrowing from the Jews—one would have to ask *why* those cultures would do so. The pagan civilizations we are speaking of were *global superpowers*, ruling the known world at that time for many thousands of years, while the Jews were *only a small nomadic tribe of shepherds* in one of their many conquered regions? Even the Old Testament writings themselves say the Israelite nomads were never more than wanderers or slaves until approximately 1500 BCE (the disputed time of the Exodus)[48] or around 1000 BCE (when the nation of Israel was born). Why would those superpowers with histories and mythologies going back from 2000 BCE to even 5000 BCE embrace the mythology of this small desert nomadic tribe as part of their own culture?

The Origin and Evolution of the Bible

Above, we explored the various cultural, political, and religious forces that were acting on the evolution of Judeo-Christian thinking and text-writing. Humans have long been responding to a divinely inspired inner urge to search for God, leading to the generation of all kinds of oral traditions and

---

48. Stone, *Ancient Israel's History*, 127–64; Dever, *Who Were the Early Israelites*, 8.

mythologies. We have convincing evidence that those oral traditions go back to at least 5000 BCE: they are evident within the roots of the Babylonian civilization, whose writings are the oldest that we have. But radiometric dating of ritualized burials suggests that some oral religious beliefs go back fifty or a hundred thousand years. Also above, we explored the concept of humans sharing and integrating their various responses to the biggest questions of life. How did we humans come to be? What does it mean to be human? What happens when we die? And many other such questions. The human authors were inspired to commit those oral traditions and mythologies to writing in various languages and scripts, and on various media (including stone, clay tablets, papyrus, animal skins, pottery, and statues). In both cases, the ideas and texts were a collaborative effort between humans and the Divine (the relative contributions of which would vary from one civilization to the next, each religion preferring to claim that their own contains the highest ratio of Divine:human content).

Now we will explore the idea of how the Bible itself—comprised of Hebrew, Aramaic, and Greek texts—came into our possession. Some readers might think this is redundant, given that we have already discussed authorship of individual texts. If they do, those readers might not have taken time to consider the making of "the Bible." They will almost certainly acknowledge that God did not lower the Bible down out of heaven in the form we now have it, but they nonetheless may not correctly understand how the different texts were collated together. It is not as though God directly told each author to "write the next book in the series," nor that once each manuscript was written it was immediately appended to a growing document until finally the Apostle John wrote his book of Revelation and concluded it with the dire warning against adding or removing anything to the final product (Rev 22:18–19). One can easily claim that God directed the collection of the manuscripts into one unified document, but, once again, we will see that this step also included a very distinctly human contribution. The fact is that thousands of texts were written by authors who each believed they were writing eternal and divinely revealed truths, but less than one hundred books were ultimately brought together into something one would call "the Bible." I have used quotations here because the word "Bible" simply comes from the Greek word for library, and different groups of people have included different texts within their most important religious "library," a process that scholars refer to as canonization. Judaism, for example, recognizes a smaller number of texts, most of them in Hebrew, including the ones that Christians call—somewhat insensitively—the "Old Testament." Christians attach to that collection of books varying numbers of other Greek texts that they refer to as the "New Testament," and may also insert between the two Testaments a variable number of "apocryphal books." For this reason, the Protestant Bible has 66 books, the Catholic Bible has 73, the Greek Orthodox Bible has 76, and the Ethiopian Orthodox Bible has 81 books.

We can begin with the Hebrew Bible. Scholarship has revealed that the Hebrew system of writing first appears approximately in the year 1500 BCE.[49] It is worth reminding the reader that a significant portion of the narrative and historical content of the Old Testament is set in a context thousands of years earlier than that, which raises the question of what oral language had been used to first carry it before it appeared in the written form(s) that we now find in the Hebrew Bible. Once Hebrew script was invented, a very large number of texts were written. For some of these we know only their titles and a very cryptic summary of their contents, since they serve as footnotes and are cited by other Old Testament authors: these include the *Annals of Samuel the Seer* (1 Chr 29:29); the *Records of Nathan the Prophet* (1 Chr 29:29; 2 Chr 9:29); the *Records of Gad the Seer* (1 Chr 29:29); the *Visions of Iddo the Seer* (2 Chr 9:29 and 12:15); the *Records of Shemaiah the Prophet* (2 Chr 12:15); the *Book of the Wars of the Lord* (Num 21:14); the *Book of Jashar* (Josh 10:13); the *Annals of the Acts of Solomon* (1 Kgs 11:41); and the *Annals of the Kings of Judah* (1 Kgs 14:29). The knowledge that there are so many titles for manuscripts now out of our possession gives us good reason to think that there likely have been many other manuscripts which are also now lost and about which we will never know anything (but which the Hebrew scholars of that era did).

On the other hand, in many other cases we have complete copies of Hebrew texts. It is important for the reader to recognize that these are only *copies* of the original texts: not the originals themselves. The fact that copies are often different from the originals (which some would naturally refer to as "errors") gives sufficient reason to soften or even question the claim of scriptural inerrancy and infallibility, two concepts which some turn into an insistence or requirement, and yet others turn into a shibboleth: a word to literally fight and die for (see Judg 12:4–6). Some believers will maneuver around this delicate matter by redefining the meaning of the words "errant" and "fallible": for example, they might refer to "apparent errors," which really does not change the problem but only eases the discomfort for those who hold a very delicate view of scriptural authority. Others do so by revising their dogmatic statements by claiming the documents "were inerrant *in their original form*," but that only moves the goalposts one step back: one can *believe* that the originals were "inerrant," but can never *know* it because

---

49. Cahill, *Gifts of the Jews*, 6; Sanders, *Invention of Hebrew*, 103; Finkelstein and Silberman, *Bible Unearthed*, 22.

we have not had the original documents for thousands of years (and have nothing to match them up against in order to confirm that there are no errors). The closest we can come in claiming access to the original wording and content is by pointing to the first official Greek translation of the Hebrew texts—the Septuagint—produced in the second and third centuries BCE using copies much older than the ones we now have (again, emphasizing that these were nonetheless *copies*, not originals). The Septuagint may be the closest to the originals that we have, but is by definition vulnerable to the distorting effects of translating from one language to another. An often quoted example of this is the disagreement over how Matthew interprets Isaiah's Hebrew word for "a young girl" (Isa 7:14) with the Greek word "a virgin" (Matt 1:23). A less familiar example that we will consider below is why the Hebrew word for "garden" (*gn-'dn*) was translated using the Greek term *paradeisos* (royal hunting park or orchard) rather than *kêpos* (garden). The books that Jewish scholars and religious leaders worked with, and that we now also have copies of, include not only those found in the Hebrew Bible or "Old Testament," but also many apocryphal books (*Ecclesiasticus*; *Tobit*; *1 and 2 Esdras*; and others) and yet others (*Gospel of Thomas*; *Shepherd of Hermas*; *Epistle to the Laodiceans*; and many others).

So out of the thousands of texts that were written, how does one determine which of them were sufficiently divinely inspired as to merit inclusion in the Hebrew Bible? That selection process—which again is referred to as canonization—is nowhere explicitly defined. There was no committee of Hebrew priests or any other sanctioned body of Hebrew scholars who officially agreed on a common list. Instead, over the course of time, certain books became increasingly recognized as bearing the stamp of Divine authority. The temple at Jerusalem (and later, local synagogues) included more and more copies of a given author's work(s) within their collection of scrolls, and unanimity of opinion began to take form: one can argue that the "opinion which took form" was in fact at the prompting of the Holy Spirit. By the time of Jesus, there was a general consensus as to which Hebrew texts should be included in what we now call the Hebrew Bible, or "Old Testament," although there may still have been differences between one synagogue and the next. Interestingly, Jesus refers to certain writings which do not appear to be found anywhere in our modern "Old Testament" (Luke 24:45–47).

In similar fashion, many of the Greek and Latin texts which were later written grew to be increasingly accepted, including those that now appear in our "New" Testament and many others which do not. The latter include: the *Acts of Thomas*; the *Revelation of Paul*; the *Apocalypse of Peter*; the *Gospel of Peter*; the *Gospel of Philip*; the *Gospel of Judas*; the *Gospel of the Nazarenes*; the *Gospel of Truth*; *Gospel of the Ebionites*; the *Epistle of Barnabas*; Paul's

Letter to the Laodiceans; *3rd Corinthians*. Biblical scholars and Christian leaders disputed frequently through correspondence and smaller gatherings regarding which books should be officially recognized or canonized, and congregated several times in large groups specifically to discuss this important point. Eventually, a consensus grew over an extended length of time regarding which books to include as Scripture and which to exclude. Although this process involved prayerful consideration—and upon that fact one might base a claim that the scriptural canon was Divinely provided and ordained—it must also be recognized that there was tremendous dispute, argument, and division over this matter: this is all evidence that the *human* element also made a significant contribution to this decision.[50] Even as recently as 1522 CE, Martin Luther—the father of the Protestant Reformation—was advocating the removal of the books of Hebrews, James, Jude, and Revelation from the canon because they did not line up with his theology.[51] Again, the reader should keep in mind that different Christian traditions include different numbers of books in their modern canon.

Conclusion

The reason I have gone through so much history regarding how religious ideas were conceptualized, orally transmitted, committed to writing, and collected into libraries ("Bibles") is to emphasize the large role that humans have always played in the entire process. Many Christians are completely unaware of this part of the important story of how we came to receive the Bible: it is generally not taught from the pulpits on Sundays nor discussed at midweek gatherings. Much is said and taught about Divine inspiration of the biblical authors, and certainly this author accepts the importance of Divine inspiration. However, irrespective of how one understands God's role in giving understanding to the church (particularly revelation, inspiration, and direction), one has to recognize that we humans have inserted ourselves at every step along the way. Much of what we hold on to dogmatically as God-given truth includes tradition handed down to us by our theological ancestors (Paul's injunction about hair length is one example; infant versus adult baptism is one of many others). This point will be important as we now explore how our understanding of the immaterial aspects of human existence or the afterlife has changed, sometimes dramatically, over the course of the past five or ten millennia.

---

50. Beckwith, "Canon," 27–34.
51. Scobie, "History of Biblical," 12.

CHAPTER 2

Premodern Human Ontology: The Mind, Soul, and Spirit

Ontology—the theory of existence and what it means to be—has been pondered for millennia. It is claimed that "thinkers have offered no less than 130 different views of human ontology,"[1] and though the Bible may rule out many of these, there still remains a large number which are all compatible with Scripture. The Bible never explicitly defines the nature of the soul or spirit, which is actually quite puzzling or even ironic, given that one major aim of Scripture is spiritual development and ultimate questions about the soul.

Broadly speaking, Christians prior to the Enlightenment were relatively unanimous in their answer to this question, pointing to the first few chapters of Genesis as their starting point and then finding other passages scattered throughout the Old Testament when and as needed. Those passages provide a certain level of detail regarding the *biological* aspects of what it means to be human, but seem to speak in only vague terms about our inner nature (our consciousness, mind, soul, and/or spirit) and our ultimate destiny (heaven, hell, and the afterlife). The Christian view(s) on those aspects of anthropogony were developed much later during the Hellenistic and Patristic periods (the first few centuries before and after Christ, respectively).

> Christian anthropology was thus mainly occupied with the twin ideas of how the human being was constituted (that is, the nature of body, soul and spirit), and how the human being had fallen from grace (the fall) and could return to God's glory.... Paul regularly talked about the tripartite division of humans into soul (*psyche*), spirit (*pneuma*), and body (*sarx, soma*). There

---

1. McFarlane, review of *The Human Person*, 94; also see Cortez, *Theological Anthropology*, 69.

is hardly a writer of the patristic era, whether in dogmatic, spiritual, or ascetic contexts, who does not speculate on what that might mean.[2]

However, Christian theology during the Patristic period did not develop in an intellectual vacuum. It was based solidly upon ancient Hebrew Scripture read through classically Greek-educated eyes and minds. Those early Christian theologians would have repeatedly heard the Greek creation/origin stories told by relatives and in public reenactments: that would certainly have shaped their emerging view on anthropogony. Furthermore, the understanding of origins held by the earlier Hebrew authors was undoubtedly influenced by the ancient Near Eastern zeitgeist of the Sumerians, Akkadians, and Egyptians.

In this chapter, we will explore briefly the evolution of anthropogony and human ontology from its pre-Hebrew and Hebrew roots through to the modern era, and will see how our modern Christian understanding of the "soul" has morphed over the millennia in response to the zeitgeist of the civilizations surrounding our theological ancestors (and that understanding continues to change even today).

Prehistoric Views of the Inner Human

At this point in time, we have no way to know for certain *what* prehistoric people believed about the invisible human essence, nor even *when* they started doing so. But it is possible to imagine and rationalize what the earliest prehistoric understanding would look like for the more contemplative among our long distant ancestors.

It would be obvious even to such a prescientific people that something changed when a person died. One moment, that person could be breathing, speaking, moving, and interacting with others around them, and a moment later they would lie motionless, never to move or respond again. The body was otherwise physically unchanged, but *something* had changed. Or perhaps something departed: something that was usually not seen or felt (although there might be occasions in which someone claimed to see or feel something). More often, the something which left might be heard: those prescientific people might associate that indefinable "something" with the breath, since one of the last things a dying body often did was let out some kind of an audible sound. A deathly illness might be accompanied by laboured breathing as the person struggled against death, and eventually

---

2. McGuckin, "Anthropology," 13.

lost that struggle and stopped breathing. In both cases, the last evidence of a person was usually their breath. Likewise, one might be startled awake gasping and heaving from a nightmare, which many preliterate societies understand to be a journey toward the world of the dead.

We may never know for sure, but it is certainly plausible that prehistoric, prescientific people entertained certain notions of an invisible, ephemeral essence which constituted the basic personhood of humans. The part which felt emotion, and remembered events and stories, and animated the body. But where did that essence reside within the physical body?

It took humans millennia to begin to associate thoughts and personality with the brain. The brain exhibits no externally evident activity which betrays its function in the body: it is otherwise a completely silent and invisible organ. No one takes literally the expression "I can hear the gears inside your head turning." You cannot feel any movement or vibrations or warmth in someone else's head, or even your own, during intense mental activity. There is in fact no light bulb that goes on inside or above your head when you suddenly realize something. There was absolutely no reason to link the brain to thought. It was impossible to make this association between brain and personality without modern medical and scientific technology. It may seem blatantly obvious to us today, but that is only because of centuries of scientific investigation into the question, key elements of which are constantly sprinkled in our everyday conversational language, even conversations we have with/around children, and from which they can pick up subtle clues about the function of our brain ("You can get that idea out of your head, son"; or "she's got a good head on her shoulders"). Just like a riddle, brainteaser, or magic trick which completely dumbfounds the first time it is encountered, but becomes so simple and child's play once it has been explained, we should not think that the association between the brain and thought would have been "just so obvious" to ancient humans.

Instead, it would have been noted millennia ago that deep emotion would be felt most acutely in the chest. During an encounter that raises the emotions and causes the thoughts to swirl—such as a deeply romantic one or being faced with a charging predator—one's chest would start pounding (the heart) and heaving more deeply and rapidly (the lungs). Particularly distressing news can often cause one to exhale audibly or gasp, and one might say that their heart sank in their chest in grief; one might describe sadness or even depression as "feeling deflated," and they might walk around with their shoulders dropped. A variety of emotions can cause one to feel tight in the throat. When one is particularly famished and feeling the various cognitive forms of distress that accompanies hunger, or has just enjoyed a big meal and is feeling particularly pleasant, or has just eaten something foul or poisonous and experiencing all forms of pain and discomfort because of that, all those feelings would be associated with the belly. There might even be sounds and movement coming from the abdomen at those times, as the intestines begin to churn and writhe. All of these observations—phenomenologically "scientific" ones to the thinkers of their era—pointed to the chest/abdomen as the place where the essential personhood resided.

The head, on the other hand, would give no indication of the activity going on inside of it. Absolutely nothing about the head that can be perceived without modern technology gives any indication that thoughts come from inside your cranium. Certainly, a serious head wound might cause one to stop exhibiting signs of intelligent thinking, but the same would be true if the heart were to be pierced or the chest opened. It would be an exceptionally rare event for a person to survive an injury to their head which produced a long-lasting change in their personality and which might prompt the more observant among them to associate the head with cognitive function: one example of exactly this happening would be Phineas Gage in 1848, who had a metal spike accidentally driven through his head and survived, but with a markedly changed personality afterward (we will encounter this individual again in chapter 3).[3]

So it should not be surprising that many ancient peoples viewed various organs within the chest as being the seat of the personality. This is referred to as phenomenological science: explanations based on our sensory perception of the thing being considered, much like referring to the sun as rising or setting over earth's horizon when it is in fact the earth that is spinning on its axis. We will see below that the people of the ancient Near East believed that the human essence ("soul") resided in one or more internal organs of their chest, but uniformly excluded the brain in this capacity. The Hebrews inherited that understanding from those surrounding nations, unless one chooses to believe that YHWH misdirected them to that erroneous understanding. And modern Western society too has continued that inheritance because our society is built upon Judeo-Christian thinking: we still say things like "I love you with all my heart" or "I've got a gut feeling about this," even though we know that cognitive emotion has nothing to do with our blood-pump or our digestive tract.

---

3. Parkes, "Vulnerability," 47–48; Ward, *Big Questions*, 149; Shults, *Reforming Theological Anthropology*, 179; Jeeves, "Neuroscience," 174; Brown, "Cognitive Contributions," 121; Green, *Body, Soul*, 82.

Ancient Mesopotamian Anthropogony

In the minds of the Sumerians (mid-fifth to late-third millennium BCE) and Akkadians (mid-third millennium BCE), the gods could be fickle and capricious, just as likely to take a course of action which humans might label as "evil" versus "good" (although no one was in a position to claim as much: the gods were laws unto themselves . . . "might makes right"). The relationship between the creator gods and their creatures was distant and cold: the gods made humans for the sole purpose of relieving themselves of hard labor (the latter had previously been the responsibility of the lesser gods, who eventually revolted against such service). The human creatures that they made (by mixing clay from the ground with the blood and flesh of a slain god)[4] were both immortal and able to reproduce, which eventually led to another problem: a population explosion and the noise pollution from human activity which disrupted the gods' peace. This was then solved by the king of the gods sending a universal flood, although one of the lesser gods motivated Ziusudra (Sumerian) or Atrahasis (Akkadian) or Ut-napishtim (neo-Babylonian) to build a boat and load it with animals to survive the deluge: that faithful and obedient human servant was alone rewarded with the gift of immortality. Otherwise, for the rest of humanity, death through old age and disease became the means by which the human population was contained.

The "flesh and blood" of the slain god conceptualized in the minds of the ancient Mesopotamians should not be equated to the physical substance of which we understand human bodies to be made. Instead, this was an immaterial substance that accounted for our life force, intelligence, ego, and self.[5] It essentially formed the human soul. As such, the ancient Mesopotamians held a view which was very similar to the relatively more modern concept of substance dualism—the idea that humans are comprised of two distinct substances, one physical (the body) and the other immaterial ("spirit")—that we will encounter when we discuss classical Greek philosophy. In fact, in the minds of the ancient Mesopotamians, there were three distinct soul substances: the *baâtu*, the *eθemmu*, and the *zaqīqu*, each with unique features and "functions"[6] (this detail will become relevant when we encounter later civilizations which also believed in a tripartite nature of the immaterial substance: the Hebrews, the classical Greeks, and Christians).

That life-force substance retained a certain relationship or quality of the god-ness from which it derived. When an individual died, that

---

4. Scurlock, "Mortal and Immortal," 77.
5. Abusch, "Ghost and God," 366–67, 370–78.
6. Scurlock, "Mortal and Immortal," 78.

component continued to exist for a while in the underworld as a much diminished "ghost" of the individual. The legend of Gilgamesh coming to terms with the death of his dear friend Enkidu is heart-wrenching.[7] His horror at seeing a worm wriggle out of the nose of his friend's cold dead body prompted an all-consuming search to find some elusive secret to immortality, including some close brushes with the underworld (he came to learn that only the gods and Ziusudra/Atrahasis/Ut-napishtim possess immortality). The "ghost" retained aspects of its individual identity within the memory of those who were still alive, but became more diminished with every passing generation; after a few generations it was forgotten and became only a vague and indistinct collective memory of one's many ancestors.[8] Otherwise, the ancient Mesopotamians, the earliest of people groups to have left us written evidence of their religious ideas, indicated no concept of resurrection, final judgment, or an afterlife coexistence with the deities.

Ancient Egyptian Anthropogony

The ancient Egyptians (late-fourth millennium to first century BCE) developed a very elaborate mythology, most of it revolving around the lives of the gods. In general, humans were only a very minor part of those stories. Their kings and Pharaohs, on the other hand, were essentially divine, "born as a seed of the creator-god, implanted in a woman, and so not physically human."[9] They saw the human personality residing within the chest rather than the cranium. As such, embalming the bodies of important individuals involved carefully storing the lungs, intestines, stomach and liver in canopic jars, and placing the heart back into the embalmed body to preserve it for judgment in the afterlife, while the brain was unceremoniously siphoned out through the nasal cavity and thrown away. Humans are mortal, but when they die, their soul enters into an afterlife in the netherworld. The heart of the deceased would be weighed and if found wanting it was dropped to the floor and eaten by the god Ammut; otherwise, the heart/soul continued on to the Field of Reeds, an eternal place of purification and bliss.

A key theme for the Egyptians, one which will be developed below when we look at the Hebrew understanding of ontology, is that divine breath is the source and sustainer of life in general, and the human soul in particular. This relationship is found in many passages of Egyptian mythology:

7. George, "Epic of Gilgamesh."

8. Toorn, *Family Religion*, 55–58; Abusch, "Ghost and God," 372, 373; Scurlock, "Mortal and Immortal," 78–79.

9. Quirke, *Exploring Religion*, 47.

- in a hymn of Akhenatan: "You give life to the children among the women and among the men who give seed, you order the child in its mother's womb . . . you give air to bring to life all that you create . . . from the moment the chick leaves the egg . . . you give it your breath to bring it to life";
- in an address to the god Amon: "Your color is light, your breath is life . . . your body is a breath of spirit for every nostril, we breathe through you in order to live";
- to the sun god: "The spirit of life and its warmth both come from your nose. . . . Eternal life is in you, to give life [by] your inspiration into the nose of living [creatures]";
- the god Re, addressing Osiris: "May the breath of Re be transferred to your nose, and the vapor exhaled from the mouth of King Khepri be close to you, so that you be able to live";
- Seti II asks the sun god: "Turn to me . . . O rising sun, who lights the lands with your glow . . . giver of breath."
- in praise of Ramses II: "He gives breath at his will; all the earth gives thanks unto him";
- in prayers lifted by the dead: "O Atum, give me the spirit that is in your nose," and "O Atum, give me the pleasant breath that is in your nostrils."

In fact, it has been proposed that this relationship between divine breath and human soul/life actually originated in ancient Egypt and then spread to all other cultures.[10] If this is true, that would be of great importance to the subject of this book, given that the Hebrews also used "breath" in descriptions of God's essence and actions, as well as those of the inner human essence.

Ancient Greek Anthropogony

The ancient Greeks and their Mycenean and Minoan ancestors (second millennium BCE) may not have influenced the ancient Hebrew authors directly, but I will present their thinking here because their Hellenic and Hellenistic descendants—including Aristotle, Socrates, and Plato—did *profoundly* influence Judeo-Christian thinking. They used two key words to refer to the immaterial aspects of human existence. *Psyche* (soul, mind,

---

10. Hehn, "Zum Problem des Geistes," 216.

spirit) contained all of one's thoughts and emotions and animated the physical body; the *pneuma* literally means "breath" (it is derived from the word *pneō*, or "to blow"), but is often translated as "soul" or "spirit" (for example, when the New Testament refers to the Holy Spirit).

The ancient Greeks developed various afterlife mythologies, in part influenced by Egyptian thinking, stories which were confused and even contradictory.[11] The *psyche* left the body upon death and entered the house of Hades (the brother of Zeus) to join others who had also died. The dead endured a gloomy, monotonous, semi-conscious existence of no return.[12] These were diminished versions of their previous selves: a shadow which had no strength, will, or even memory of their previous life, except occasionally when given a sip of sacrificial blood offered symbolically by feasting relatives and loved ones. It was believed that the dead interacted with the living during festivals (where they received sacrifices, and sent back good things) and when they were conjured up or appeared in dreams or visions (to convey wisdom, instruction or warnings). This is what happened when Odysseus revived the dead soul of Achilles, who in turn wailed that it would be better to be a servant of a nameless peasant in the land of the living than to be lord over all the dead.

Although certain Greek myths also refer to Hades and to the Tartaros (a fearful pit and place of punishment for certain gods and humans who were guilty of particularly heinous crimes), and also to Elysium and the Elysian Fields (a blissful world for a few select individuals who avoid death),[13] there is no concept of a final judgment or sentencing of most humans to some eternal afterlife existence, and certainly no concept of resurrection.

Ancient Hebrew Anthropogony and Ontology

There is much debate as to when the various Old Testament texts were written (and by whom); this was covered briefly in chapter 1 above. To place this next section within the historical context laid out in the sections above (that is, relative to the Mesopotamian, Egyptian, and Greek civilizations), it is enough to remind the reader that Abraham is said to have lived during the beginning of the second millennium BCE, and that the people who a millennium later became the nation of Israel were thus situated in a place and time well within the philosophical/theological incubator of the ancient Near East, and under

---

11. Burkert, *Greek Religion*, 190–99.
12. Endsjø, *Greek Resurrection*, 24–25.
13. Endsjø, *Greek Resurrection*, 26; Scharen, "Gehenna," 327.

the influence of the global superpowers that had already controlled the region politically, culturally, and ideologically *for several millennia*.

The Old Testament texts give a somewhat detailed description of the material aspect of humanity (our physical body), but they lay out in only vague terms the creation and nature of our inner being. Broadly speaking, those texts describe the first human body being created by God from dirt and then infused with an ineffable essence—"the breath of life" (Gen 2:7)[14]—and the second human body being re-formed from an extracted physical portion of that first one (Gen 2:21–22). Nothing is written about how that second human—Eve, the first woman—received a separate and distinct inner essence. Doubtless she would have had one, but the text does not tell us whether it was breathed into her, or was inherent within the piece of flesh taken from Adam. Nor is anything written about how all other humans since Adam and Eve acquired their own material and inner natures, beyond vague allusions to knitting needles (Ps 139:13). For example, whether souls are preexistent before conception and are assigned to a given body, or whether they appear at some time during or after conception, the Bible does not describe in any detail whatsoever beyond merely stating it as a matter of fact: "God forms the spirit within a person" (Zech 12:1).

It is important to emphasize that the "breath of life" which God breathed into the first human(s) does not refer to the human spirit: it was not that God simply fashioned the physical body out of clay and then breathed a spirit into that body, since the animals are equally said to have the breath of life (Gen 1:30; 6:17; 7:15, 22), and no distinction is made between the breath that humans received and that which animals received. *Life itself*—human and animal—is indeed connected to that breath, and physical death is connected to removal of it (Deut 20:16; Josh 10:40; 11:11; 1 Kgs 15:29; Ps 104:29; Ezek 37:5–6); a dying person is sometimes said to have "breathed their last" (Gen 25:8, 17; 35:18, 29; 49:33; the writers of the Gospels carry this Jewish imagery into their Greek writing in Mark 15:37 and Luke 23:46).[15] The Hebrews also connected life with the blood of the person/animal (Gen 9:5; Lev 17:11, 14; Deut 12:23; Jer 2:34), much as was the case in Babylonian mythology—especially in the creation of humans in part from the blood of a slain god—as Abusch explores in detail.[16] Neither should one think that the life-giving breath which God breathed into the clay was his own spirit: in Ezekiel's vision of the Valley of Dry Bones, "the

---

14. Recall the point made above that the Egyptian civilization may have been the first to find a link between "breath" and divinity; see page 37.

15. Kosman, "Breath, Kiss," 96–124.

16. Abusch, "Ghost and God," 364–72, 376–82.

spirit of the Lord is portrayed as if it were an autonomous entity, which the prophet is commanded to invoke to action, as if it were not actually the breath of the Lord."[17] YHWH's breath is also portrayed as being behind other aspects of creation (given that God speaks Creation into existence in the Genesis narrative; also see Ps 33:6) and aspects of the weather (Exod 15:8 [compare this with Exod 14:21]; Job 37:9–10). Instead, Genesis says that this life-giving breath and creative force transformed the lump of clay into a spiritual being: it does not say that Adam was *given* a living spirit, but that he *became* a living spirit.

The ancient Hebrews had their own equivalent of the Greek word *pneuma*: for them, *ruach* could mean "breath," "wind," or "spirit." Ezekiel's vision of the Valley of Dry Bones provides an interesting interplay between these three meanings (especially Ezek 37:9). Also like the ancient Greeks and Egyptians, the ancient Hebrews believed that our essential inner self resides in our chest and abdomen. The wording for one's bowels or intestines referred to one's deepest, innermost being. When David wrote in Psalm 40:8, "your law is in my heart," the Hebrew text is literally written, "your Torah is in my guts," and the writer of Proverbs uses similar word imagery (Prov 20:27; 22:18). The Apostle Paul carries this imagery forward into his Greek writing in Philippians 1:8, where he uses wording that the King James translates, "I long after you all in the bowels of Jesus Christ." The bowels (and the liver) are seen as the seat of emotion (Lam 2:11).

In contrast to our own modern understanding, the Hebrew texts *never* refer to the brain as the organ which confers cognitive ability. The exterior of the head itself can wear or display honor, shame, or guilt, but is never implied to take an active role in cognitive function:[18] for example, the head is anointed with oil, or wears a crown; the Nazarite vow forbade the cutting of one's hair (Num 6:5; Judg 13:5), and Samson's distinctive strength was linked to his hair (Judg 16:17); the Apostle Paul (writing in Greek, but thinking like a Jew) attributed a great deal of importance to the honor inherent within the hair length of men and women (1 Cor 11:14–15); the head bears the responsibility of a crime (Ezek 9:10; 11:21; 16:43). But what is *inside* the head—the brain itself—is *never* linked to personality, will, honor, or guilt: in fact, there is not even a word in the Hebrew Bible for "brain."

Dr. R. D. Branson has provided a very enlightening review of biblical references to three specific organs within the human body. He was especially interested in their correspondence with cognitive function and whether those biblical references indicate that the ancients had an accurate understanding,

---

17. Kosman, "Breath, Kiss," 104.
18. Pedersen, *Israel*, 174.

which could be claimed to have been given to them by YHWH, of the physiological functions of the three different organs in particular or even of human physiology in general.[19] He finds that kidneys are referred to eleven times in the Old Testament, but *never* in the context of filtering blood and removing metabolic wastes. Instead, the closest English words used in the various modern biblical translations for the Hebrew understanding of the kidneys are "heart," "inmost being," "soul," and "mind." The organ that we now call the heart is referred to 853 times in the Old Testament and 148 times in the New Testament, but *never* in the context of the function that we now recognize it to have: pumping blood throughout the body. Instead, all of these biblical references to the heart pertain to a wide variety of feelings, emotions, or states of mind, including joy, grief, courage, fear, desire, lust, understanding, thinking, and inner reflection; the heart, *not the brain*, is singularly connected to our relationship with God.

Branson addresses the possibility that the Hebrews only intended these words for "kidney" and "heart" to be taken metaphorically, much as we do today (again, speaking of loving "with all one's heart" or "having a gut feeling" about something). But he then makes the following crucial observation:

> It would be helpful, however, if the Bible gave the metaphor's referent, but it does not. In over a thousand usages, there is not one instance in which the word for heart or kidneys refers to or describes their physical functions. While the Israelites were aware that the head did house an organ, there is no word in the Hebrew Bible for the brain. This lack of knowledge of the function of the brain is in keeping with the other cultures in the Ancient Near East.[20]

This would therefore be another example of the biblical texts not advancing knowledge beyond the science of the period in which they were written;[21] and in this sense, it is also an example of the human contribution to the authorship of the biblical texts sometimes eclipsing that of the Divine (since I doubt that any reader will be adamant that God himself sees the intestines as the seat of emotion, nor that the brain is irrelevant to cognitive function).

The biblical authors used several words to refer to the central essence of a person, including those words which are most commonly translated

---

19. Branson, "Science, the Bible, and Human Anatomy," 229–36.
20. Branson, "Science, the Bible, and Human Anatomy," 231.
21. Walton, *Lost World of Genesis*, 17, 19; Lamoureux, *Evolution*, 85–112.

as soul (*nephesh*; used 754 times[22]), spirit (*ruach*; used 361 times[23]), and heart (*lebh*; used 853 times[24]).[25] The meanings captured within these three words are not identical but overlap significantly, and thus are often used interchangeably. But they are nonetheless quite distinct.

The "heart"—*lebh*—referred to one's essential personhood and character: David was a man after God's own *lebh* (1 Sam 13:14). It was the seat of volition (dozens of passages refer to the hardening or softening of someone's heart), courage ("their hearts melted within them"; Deut 1:28; Josh 5:1; 7:5; 14:8; 2 Sam 17:10; Ps 107:26), and memory (Pedersen elaborates on how the word "remember" can be literally rendered "to rise up from one's heart").[26]

And yet *nephesh* is a word used in all Semitic languages (albeit in slightly different forms) to refer to the whole of a person where *lebh* or *ruach* do not. *Nephesh* referred fundamentally to breath, and metaphorically to desire, but certainly to a living embodied personhood.[27] Jacob came to Canaan with seventy *nephesh*, not seventy hearts or seventy spirits (Gen 46:27; Exod 1:5). When Abram and the king of Sodom were dividing up the spoils of war, the latter said, "Give me the *nephesh*, and you keep the possessions/goods" (Gen 14:21). When a census is taken, the question always asked is "how many souls are there?" And yet interestingly, the word *nephesh* can also be used in referring to animals (Gen 9:10; Prov 12:10) or even God himself (Judg 10:16; Isa 1:14).[28]

Many modern scholars emphasize that the ancient Hebrews held a holistic view of human existence, rather than a strictly dualist one in which an immaterial soul conveys the actual personhood and the material body is merely a vessel.[29] The word for that totality is *nephesh*. And yet, the ancient Hebrews were still able to see the two components distinctly: Isaiah refers to the wasting away of both *basar* (flesh, meat) and *nephesh* (Isa 10:18). David writes that his *nephesh* thirsts and his *basar* longs for the living God (Ps 63:1). Job refers to the pain felt by the *basar*, and the mourning expressed

---

22. Moreland and Rae, *Body and Soul*, 27.

23. Moreland and Rae, *Body and Soul*, 30.

24. Branson, "Science, Bible, and Human Anatomy," 230; also see Waltke, *Book of Proverbs*, 90.

25. Pedersen, *Israel*, 102; Allen, "Persons," 166; Ward, *Big Questions*, 138; Green, "Restoring the Human Person," 4; Green, *Body, Soul*, 54–60; Waltke, *Book of Proverbs*, 89–93.

26. Pedersen, *Israel*, 106–7.

27. Dr. Gus Konkel, personal communication.

28. This word usage can be seen in versions such as the Orthodox Jewish Bible, but is removed in other standard English translations of the Bible.

29. Pedersen, *Israel*, 99.

by the *nephesh* (Job 14:22). Interestingly, heart (*lebh*) and spirit (*ruach*) could both refer to the core nature of a person, and yet the Hebrews might refer to them separately: David asks God to create in him a pure heart, and to renew a steadfast spirit (Ps 51:10), while Ezekiel prophesies that God will give Israel a new heart and a new spirit (Ezek 11:19; 36:26). The *lebh* could be changed (Ps 51:10), or exchanged (Ezek 11:19; 36:26). The spirit could be distributed to someone else (Num 11:17–25; 27:18–20; 2 Kgs 2:9–15), or be transported far away from the body in dreams and visions (2 Kgs 5:26; Ezek 8:3; Dan 8:2). We will encounter this ambiguity again later below, when we see how the Hebrew view is radically revised under Hellenistic influences, and developed even further when Christianity takes those Hellenized Hebrew ideas and thoroughly infuses them with classical Greek philosophy during the Patristic era.

It seems that the ancient Hebrews believed the first humans were created mortal. For example, the fact that God provided the pre-fall humans with plants and fruit as food (Gen 1:29), and after the fall found it necessary to deprive them of the Tree of Life in order that they would not live forever (Gen 3:22–24), is interpreted by some as indications that the first humans had always been subject to decay and death and that the Tree was given as a remedy or antidote against those: one would not be given a Tree of Life if one were already immortal.[30] Besides, Genesis describes God creating humans and then commanding them to "be fruitful and multiply" (Gen 1:28): all this reproduction over countless millennia without any dying would have produced a seething mass of human flesh that would have totally covered the face of the earth to inestimable depth. We have no fossil evidence of that kind of human population explosion.

This assertion of original mortality does not necessarily conflict with the later divine warning that disobedience would bring death, since being driven from the garden and deprived of the Tree of Life would be equivalent to a death sentence. This is an important point. It can mean that death was not necessarily a punishment for disobedience but rather a consequence, much like a parent warning a child that "if you play with matches, you're going to get burnt" (the parent will play no role whatsoever in that foretold injury). The Creation narratives contain nothing to support the idea that humans were created immortal. According to the *Jewish Encyclopedia*:

> The belief that the soul continues in existence after the dissolution of the body is . . . speculation . . . nowhere expressly taught in Holy Scripture. . . . The belief in the immortality of the soul

---

30. Harlow, "After Adam," 188; Madigan and Levenson, *Resurrection*, 88–89; Walton, "Lost World of Adam," 73–74.

came to the Jews from contact with Greek thought and chiefly through the philosophy of Plato, its principal exponent, who was led to it through Orphic and Eleusinian mysteries in which Babylonian and Egyptian views were strangely blended.[31]

The ancient Hebrew texts are not clear about what does happen after death. The author of Ecclesiastes writes that "the *ruach* returns to God who gave it" (Eccl 12:7); Ezekiel describes a vision of a large Hebrew army (which was explained within the vision to be a metaphor for the people of Israel) being resurrected from the dead when the *ruach* returned to their dead bodies. When Elijah raised to life the son of the widow at Zarephath, the text says that the boy's *nephesh* returned (1 Kgs 17:21–22). When the medium at Endor answered King Saul's request to bring back the spirit of the prophet Samuel, she used very different wording—*elohim*—which is elsewhere used to refer to disembodied nonhuman spirits (1 Sam 28:13). Otherwise, we will find in chapter 4 below that the ancient Hebrews generally believed in a postmortem existence in Sheol, a gloomy land of no return characterized by dust and silence, and that this belief was later developed into the concepts of heaven and hell.

Before leaving the ancient Hebrew understanding of human ontology, we will touch briefly on one somewhat peripheral matter: the Nephilim. The author(s) of Genesis inserts a puzzling vignette shortly after the Creation account in which the "sons of God" were having children with the daughters of men (Gen 6:1–4). The latter are clearly referred to as being human: in Hebrew, *banot HaAdam*, which English versions all render as "the daughters of men," or as "human women," or some similar variant. But the Hebrew word used to indicate the fathers in this strange union is *elohim*, the same word discussed above (page 5) to refer to the God of Israel, to the pagan gods of other nations, to angels, and to the disembodied spirits of dead humans. The progeny of this strange union are not ordinary humans: they are defined in this passage of Genesis as "the heroes of old, men of renown," and elsewhere as giants (Num 13:33).

This passage raises all kinds of questions for a modern thinker. Are these "sons of God" intended to be immaterial, spiritual beings? If so, how would they have all the complementary chromosomes necessary to fertilize the eggs of the daughters of men? It is not my intention here to challenge the doctrine of the Immaculate Conception and the Virgin Birth, since one can claim that if God can create a universe *ex nihilo*, then surely fertilization of an egg would not be an insurmountable barrier for him. However, this strange story in Genesis revolves around beings who are implied to be

---

31. Adler, *Jewish Encyclopedia*, "Immortality of the Soul," 564–66.

somehow evil: would such lesser spiritual beings also be supernaturally capable of remarkable creative acts? How did the authors envision these half-human beings surviving Noah's flood (Num 13:33)?

Contemporary scholarship provides a possible solution to the modern exegete dealing with this conundrum. It is proposed that "the Sons of God" were the sons of Seth, who "began to call upon the name of YHWH" (Gen 4:25-26), and "the daughters of men" were the descendants of Cain who did not subordinate themselves to YHWH.[32] At issue here, then, was not a transgression of barriers between the divine/supernatural and the human/natural, but rather one of following versus rebelling against YHWH: the human ancestral line which YHWH had sanctified (set apart) to carry his message of salvation was being diluted and corrupted by nonbelievers or even worshippers of other gods. This not only dispels the oddity of a mating between spiritual and human partners, but also explains the reappearance of "the Nephilim" after Noah's flood (Num 13:33): these would now be the sons of Shem and the cursed sons of Ham (Gen 9:22-27), who also took divergent paths in their worship of YHWH but also began to intermarry despite YHWH's clear prohibition against that. But such an exegesis only raises other questions which take us far outside the scope of this book.

Irrespective of how one interprets "sons of God" in this passage, other questions still remain. Why did this union prompt God to proclaim in apparent exasperation: "My Spirit will not contend with humans forever, for they are mortal; their days will be a hundred and twenty years" (Gen 6:3)? How did humans contend with God in a way that could be solved by limiting their life span? Or how did limiting our life span to a hundred and twenty years solve the problem of immaterial beings mating with the daughters of men, or the "problem" of humans increasing in number, given that humans can be reproductively fertile in their early teen years and continue to be so for many decades (even into their centenary years for males)? One interpretation of this curious passage is that the "one hundred and twenty years" does not refer to the life span of individual humans, but to the time remaining for the human race as a whole:[33] the countdown for the coming global flood which would destroy them had been started.

Classical Greek Anthropogony and Ontology

This summary of a new strand of human philosophy will stand quite apart qualitatively from the previous summaries: the attention to minute detail and

32. Waltke, *Genesis*, 115-16.
33. Waltke, *Genesis*, 117.

the efforts made to explain diverse facets of human ontology are testament to the immense time and effort paid by Hellenic thinkers to such questions. I would ask the reader to pay attention to those details—and keep a mental inventory of them and how they distinguish themselves from ancient Babylonian, Egyptian, and Hebrew thinking—in order to be more aware of how those details make their way into Hellenistic Hebrew theology, which is then in turn dramatically transformed into Christian thinking.

In the fifth century BCE, pre-Socratic thought focused on the essence or substance of things[34]—that which stays the same beneath the change around us—while Plato (428–347 BCE) proposed a metaphysical dualism between a celestial realm of eternal unchanging Forms or Ideas and another earthly one of temporal change.[35] "God" *must* be perfect, rational, good, unchanging, and nonmaterial.[36] Whereas other substances were either material or immaterial, Plato saw humans as a combination of both. He envisioned an eternal Form or Idea of humanity, a universal immaterial substance (the "world soul," or "soul of the all," or the *nous*) from which all human souls derive their being, and the immaterial soul becoming trapped within a material body. The soul, then, strives to be freed from the material body in which it is imprisoned, and to return to the realm of the Forms from which it had fallen to rejoin the eternal Form or Idea.[37] That striving was a lifelong struggle, and was fully realized or achieved through death.

Where Plato saw a strong dichotomy between body and soul, Aristotle (384–322 BCE) saw the human soul itself as the Form of the body (rather than being an emanation of the Eternal Form which was trapped in the body), so the two could not be so easily separated.[38] He did not distinguish here between material and immaterial, matter and non-matter: instead, he still saw the soul as being "constructed from elements of the world" and occupying space.[39] We will see later that Christian theologians in the Patristic era—thoroughly Greek in their education—would navigate between these two poles and eventually settle upon a more Platonic (actually, Neoplatonic) understanding of the human soul.

Plato and Aristotle also squared off differently in their understanding of how the soul itself was structured. Both saw the latter to have three

---

34. Shields, "Philosophy before Socrates," 1–34.

35. Shults, *Reforming Theological Anthropology*, 12–13; Dillon, "Origen and Plotinus," 9.

36. Larson, *Understanding*, 69.

37. Hochschild, *Memory*, 166; Shults, *Reforming Theological Anthropology*, 166.

38. Shults, *Reforming Theological Anthropology*, 167.

39. Green, "Bodies," 160; Green, *Body, Soul*, 52.

distinct parts (as did the ancient Babylonians: see page 35), but labeled them differently. Plato referred to the rational aspect (*logistikon*), the spirited aspect (*thymikon*), and the appetitive aspect (*epithymetikon*), but Aristotle divided these into the vegetable (nutritive) soul, the animal (sensitive) soul, and human (rational) soul.

Epicurus (341–270 BCE) philosophized at length about the composition of matter, and introduced the concept of the smallest conceivable unit of matter from which every physical thing derives: the atom (the word literally means "indivisible").[40] Atoms could be neither created nor destroyed, but could be rearranged in different configurations, accounting for the diversity of things all around us including the body and the mind/soul. When the body dies, the soul dies with it: as such, there is no afterlife, and therefore no judgment of the soul.

Stoicism (third century BCE till third century CE) held that the universe is a reasoning substance which acts on matter; Stoics referred to it as the *Logos*, as God, or as Nature.[41] Humans possess souls which are emanations from the *Logos* and find happiness in part by learning how to control passions (fear; anger; lust), apply reason, and practice a high form of ethics which brings them into alignment with the *Logos*.[42]

Neoplatonism (a revised or updated version of Platonic thinking; third to sixth centuries CE), on the other hand, served as the bridge between Hellenism and Christianity, providing the latter with ready-made structures for religious thoughts, and orienting it toward the ways of looking inward from which these structures had been created. As Camus put it: "Alexandrian thought moved Christianity to reconcile metaphysics and primitive faith" (by the latter, he meant Judaism).[43] Neoplatonists built on Plato's ideas of Forms and Ideas, referring to an immaterial, unknowable creative force ("the One," "the All," or "the Good") which exudes a perfect image of itself (the *nous*). The *nous* derives from "the One" and yet is distinct from "the One";[44] the *nous* in turn serves as the archetype of all existing things and ultimately connects material things, including humans, to "the One." According to Plotinus, intellect is a procession and image of "the One," and soul is analogously a procession and image of intellect.[45]

---

40. Tate, *Biblical Interpretation*, 58–59.

41. Long, *Hellenistic Philosophy*, 118–209; Dillon, "Origen and Plotinus," 10.

42. Long, *Hellenistic Philosophy*, 197–99; Shields, *Ancient Philosophy*, 150–54, 169–71, 182–207.

43. Camus, *Christian Metaphysics*, 111.

44. Lloyd, *Anatomy of Neoplatonism*, 126; Hochschild, *Memory*, 47–48.

45. See Plotinus, *Enneads*, 3.5.9.19, 5.1.3.9, 5.1.6.45.

Humans must work to find their way back to "the One" through various virtues and ethical practices including contemplation and meditation, a process referred to as "reversion" (in distinction from the modern Christian word "conversion" because it was seen to be a *returning to* something not a *turning away* from something).[46]

Even up until the second century CE, Greek philosophers debated over the anatomical location of intellectual function, some finding that to be the heart (Aristotelians and Stoics) while others pointed to the brain (followers of Plato and Hippocrates). It was only when the Greek anatomist Galen (130–217 CE) performed live dissections on animals that a definitive connection was made between the brain and functions such as control of the body and intellectual activity. However, it remained entirely unclear how the mind and brain exerted their functions upon the body until the Renaissance era, as we will see later (page 61).

In conclusion, Greek philosophy at the time of the writing of the New Testament was not monolithic with respect to human ontology.[47] Contrary to what may otherwise seem to be the case, it is an oversimplification to say that "the Greeks" saw humans as a duality of an immaterial soul and a material body: there was instead a range of views regarding precisely how material the soul was.

Hellenistic Jewish Anthropogony and Ontology

The Hebrew understanding of human ontology, especially the soulish aspect of our existence, developed markedly in response to classical Greek thinking: many of the strands of Greek philosophy presented above were woven into Jewish (and Christian) theology.[48] This development can be seen clearly in many apocryphal and pseudepigraphic texts (the *Book of Jubilees*, *Sibylline Oracles*, *2 Enoch*, and *Tobit* in particular address the creation of humans, the fall and the afterlife), and in the writings of Philo. A Hellenized version of Hebrew thinking was the one with which Jesus, the disciples, and the early Christian church operated: the New Testament texts may have been written in Greek, but it was done so by people who were Jewish in their thinking. For them, there could be some kind of vague distinction between "soul" and "spirit." These could be overlapping in meaning, almost synonymous and yet meriting separate mention, as in Mary's Song (Luke 1:46–47). But Paul uses these words in a way that suggests a

---

46. Lloyd, *Anatomy of Neoplatonism*, 126; Hochschild, *Memory*, 49.
47. Green, "Bodies," 159–63.
48. Green, "Bodies," 161.

bit more distinctiveness, possibly as distinct from each other as they are from the body: "May your whole spirit, soul and body be kept blameless" (1 Thess 5:23). The writer of Hebrews widens the separation between them even more when describing the "word of God [as] alive and active. Sharper than any double-edged sword, it penetrates *even to the dividing of soul and spirit*" (Heb 4:12; emphasis added). Was their ability to distinguish between these two very similar terms heightened in a society which knew of Plato's and Aristotle's references to the three parts of the human soul? Whether or not that was the case, they *did* seem to be more sensitive to some kind of divisions within our inner being.

In that world of deep philosophical thinking, a peculiarity in the wording of the Hebrew text began to draw their attention to the nature of the first created humans. While many biblical passages refer to the creation of man, the first two chapters of Genesis are the only ones in the entire Bible which describe the creation of woman. This is not merely a point of minutia or fodder for conversations around gender issues. It is instead a foundation on which several elaborate strands of theology were later developed by the fathers of the church in the Patristic era, based on subtle grammatical peculiarities which are easily missed in a casual or overly familiar reading of the passages. In particular, Genesis 1:27 switches provocatively from the singular "adam" to the plural. That is, the phrase "in the image of God he created *him*" is immediately followed by the plural "*male and female* he created *them*."[49] Likewise, the fifth chapter of Genesis begins with the singular ("On the day that God created man he made *him* in the likeness of God") and then switches immediately to the plural ("he created *them male and female*; he blessed *them* and called *them* adam on the day *they* were created"). Furthermore, while the first and fifth chapters of Genesis seem to describe the creation of man and woman as a single event, the second chapter of Genesis clearly depicts first an individual being created from dust outside of the garden, and *only later* being placed into the garden and *even later yet* essentially being divided in half through a very different kind of act (not mere manipulation of clay) to produce two different beings, one distinctly female and the other male (quite reminiscent of the Greek myth introduced above of Zeus dividing primeval humans in half).

---

49. From the Complete Jewish Bible, with emphases added. Similarly, the Creator is first plural ("Let *us* make humankind in *our* image") and then immediately singular ("So God created humankind in *his* own image, in the image of God *he* created"). Karl Barth interprets this plurality as a premonition or foreshadowing of the Christian doctrine of the Trinity (see Childs, "Biblical Theology," 568), although other scholars deny or refute this (see Green, "Bodies," 156).

These three biblical passages were then taken together by some Hellenistic Jewish thinkers and interpreted as the original image-bearer being androgynous, asexual, and immortal, like the angels, with the potential of being male and/or female (Gen 1:27a) *before* being put in the garden of Eden, and *then later* being separated into two opposite but complementary sexual (and now mortal) beings in the garden (Gen 1:27b and Gen 2:15–22).[50]

Writing in first-century Hellenistic Alexandria, Philo interpreted Genesis on a verse-by-verse basis in two major works,[51] employing Jewish allegorical interpretation with a decidedly Platonic spin.[52] He equated the personal God of Judaism and the supreme divine principle of Platonism—"the One"—both seen as an undivided unity. Humans, on the other hand, are not undivided (as was implied in the Hebrew word *nephesh*), but are a duality of body and soul. The soul, in turn, is also divided into parts: a divine, rational upper part oriented toward "the One" (the "*nous*") and a mortal, irrational lower part(s) oriented toward the material world.[53] Philo saw Gen 1:26 describing the creation of an undivided concept of human being (a sort of Idea or Form) and Gen 1:27 pronouncing that they were neither male nor female (note: his translation is thus opposite to the rendering found in the Septuagint).[54] These non-gendered souls were then given bodily form, complete with distinct genders, in Gen 2:8 and onward. In describing the fall, Philo writes how sexual love brings back together the two halves of the original androgynous human, created after the image of God, and "sets up a desire for bodily pleasure, which is the root of wrong and of mortality"[55] (again echoing the Greek story of the primeval humans).

Also in the first century, the Hellenistic Jewish historian "Josephus also held to the independent existence of the soul, destined for immortality,"[56] and "catered to the Greco-Roman intelligentsia, formulating a body-soul dualism quite at odds with Israel's Scriptures but very much at home in the Platonic tradition."[57]

Judaism would continue to develop into two very different theological strands during the first century CE. One would become rabbinical Judaism

---

50. Noort, "Creation of Man and Woman," 4–8; Benjamins, "Keeping Marriage Out," 95; Teugels, "Creation of the Human," 108–13.

51. Philo, *Legum Allegoriae*; Philo, *Questions and Answers on Genesis*.

52. Hoek, "Endowed with Reason," 63–64; Green, "Bodies," 162.

53. Hoek, "Endowed with Reason," 65–66.

54. Hoek, "Endowed with Reason," 68.

55. Benjamins, "Keeping Marriage," 95.

56. Green, "Bodies," 162.

57. Green, *Body, Soul*, 26.

following the destruction of the temple in Jerusalem, and would devise yet other remarkable claims about anthropogony (for example, the Talmud claims there were 974 generations of humans before God created Adam).[58] The other strand would develop into Christianity in response to the teachings of Jesus, his disciples and apostles, and the fathers of the church.

Patristic Christian Anthropogony and Ontology

It is not hard to see how both Stoicism and Neoplatonism would resonate with, and be attracted to, the ancient Hebrew Scriptures and Christian ideas.[59] Clement of Alexandria (ca. 150–215 CE) saw Christianity as encompassing Judaism and Greek thinking, and Christians as a "third race" (the progeny of the marriage between Judaism and Greek philosophy). This Christian thinker was not simply trying to exploit Platonic philosophy as a convenient common ground with his Greek philosophical audience, in the manner that Paul seems to have done when using the Greek "statue to the unknown god" as his platform or bridgehead to the same kind of audience. Instead, Clement was convinced that all truth must cohere, and as such the truth to be found in Plato must resonate with and complement the truth to be found in Moses (the Hebrew texts) and the truth to be found in Christ. "According to him, philosophy was given to the Greeks just as the Law was given to the Jews. Both have the purpose of leading to the ultimate truth, now revealed in Christ."[60]

In the process though, those who were trained in these lines of Greek thinking would quite easily allow their Greek philosophies to influence the development of their Christian theology and cause them, for better or worse, to depart from the ancient Hebrew thinking. In some cases, these departures would be labeled heresies, while others would be accepted as orthodoxy.

For example, while Genesis taught that God created all things as good and meant to be enjoyed, they would struggle with the Greek concept that matter is evil, and sensual pleasure (the engagement of our immaterial mind with that physical matter) must be suppressed. Marcion went so far as to teach that the visible, material world was created by an inferior, evil, creator-god, and therefore all matter is evil, including the human body and its natural activities. Procreation was particularly evil, not only for its link

---

58. Slifkin, *Challenge of Creation*, 341–42.

59. McGrath, *Science of God*, 18–20.

60. González, *Story of Christianity*; also see Ashwin-Siejkowski, *Clement of Alexandria*.

to sensual pleasure, but also because it led to an increase in the amount of matter and human bodies in the world.[61] His view directly contradicted both YHWH's declaration that the creation of humans was "very good" and the Divine command to "be fruitful and multiply": Marcion resolved this difficulty by teaching that the Hebrew Scripture and the God described in Genesis were degenerate and should be discarded.

As noted above, the Hebrews seemed to believe that humans were created mortal and were destined for nonexistence in Sheol, while the Greek-thinking Christians began to see humans as originally eternal heavenly spirits, now trapped in physical bodies and seeking to be liberated in order to rejoin the immaterial.

The Hellenized Christian believers also subverted the concept of the *imago Dei*. The ancient Hebrews saw this referring to a *communal* role for all humans (male and female) as God's representatives to the rest of creation and having the God-given ability to create, subdue, and care for creation. The Greek-thinking fathers of the church, however, revised it to now refer to aspects of an individual's mind—such as reasoning, volition, and creativity—which *individuals* (some said only males)[62] can possess or acquire.[63]

As outlined above (under Philo), their Greek thinking contributed directly to the idea that humans were initially created as immortal and androgynous beings, and later re-formed into the "Adam and Eve" more familiar to Christians today. Greek thinking would also facilitate the transition from the exclusively monotheistic Judaism to a belief in the Trinity, the divinization of a human (Jesus), and heaven and hell: although modern scholars may point to hints of those concepts in Hebrew Scripture, none of them are tenets of Judaism, and some are even anathema to it.

More apropos to the central theme of this discussion, though, the fathers of the church poured considerable effort into the question of human ontology. They were quite content with the ancient Greek and Hebrew explanations that humans were divinely created from mud/dirt/clay—in an event completely separate from the creation of the cosmos, plants, and animals—though they replaced the Greek deities with YHWH/God.[64] They did, however, spend a great deal of time pondering and debating the origin, nature, and destiny of the human soul or spirit.

---

61. Stanton, *The Gospels*, 124–25.
62. Cortez, *Theological Anthropology*, 16.
63. Middleton, *Liberating Image*; Shults, *Reforming Anthropological Theology*, 217–42; Janssen, *Standing on Shoulders*, 204–15; Cortez, *Theological Anthropology*, 14–40.
64. Steenberg, *Of God and Man*, 22–9, 37, 67.

PREMODERN HUMAN ONTOLOGY: THE MIND, SOUL, AND SPIRIT

The earliest of the church fathers, raised on Stoicism which called them to get rid of all sensual passions, gladly borrowed Philo's Hellenistic Jewish interpretation of the duality of human nature—mind/reason (*nous*) and flesh/body (*sarx*).[65] They took this Hellenistic Jewish interpretation and blended it with their Stoic/Neoplatonic thinking to again produce something entirely new and different from Judaism or from Greek philosophy, as will be expounded in more detail below (parenthetically, this view of these church fathers also stands in contrast to that held by many modern evangelical Christians today).

In the second century, Irenaeus of Lyons taught that humans are bipartite in composition—comprising body and soul—but tripartite in actualization: the created body-soul was animated by God's spirit breathed into them on the first day of creation.[66] He compared and contrasted the first Adam and Eve against the "second Adam" (Jesus Christ) and Mary, particularly with respect to how the former pair failed to obey God whereas the latter pair succeeded.[67] All humans are created imperfect in the sense of being unfinished or incomplete, yet are "good" because they are oriented toward becoming perfect. Humans bear the *image* of God and yet share in what Augustine would later call original sin ("because all are implicated in the first-formation of Adam, we were bound to death through the disobedience"),[68] and so have to work to regain the *likeness* through reversion or spiritual regeneration which brings us into conformity with the image given in Christ.[69] This process of maturation—referred to as *theosis*—begins upon receipt of the Holy Spirit and grows over time in fellowship with fellow believers.[70] In the process, both the human body and its spirit would be saved and become a spiritual person made perfect at the final Resurrection.[71]

Origen (185–254 CE) revised the Neoplatonic idea—that the mind emanates from the *Logos*—into the Christian concept that it is created *de novo* by God before the body existed.[72] Furthermore, he also taught that humans are immortal spiritual beings, created without gender (as per the first chapter of Genesis) and with the ability for reason and free will (the

---

65. McGuckin, "Anthropology," 14; Heine, "Origen," 190.

66. Steenburg, *God and Man*, 38–41, 52.

67. Shults, *Reforming Anthropological Theology*, 193.

68. Steenburg, *God and Man*, 46.

69. Irenaeus, *Adversus haereses* 5.6.1; McGuckin, *Westminster Handbook*, 14; Shultz, *Reforming Anthropological Theology*, 222.

70. McGuckin, "Christ," 263; Steenburg, *God and Man*, 40–43, 52.

71. Bingham, "Irenaeus," 147–50; Steenburg, *God and Man*, 56.

72. Shults, *Reforming Anthropological Theology*, 223; Murphy, introduction to *Neuroscience and the Person*, iv.

Hellenistic understanding of the *imago Dei*), and that their free choice to disobey God led to the introduction of sin and death. As a result of this rebellion, the spiritual beings were given physical bodies (as per Gen 2) as "a place for repentance and education, providing them with an opportunity to return to their former state."[73] That body—the *sarx*—is inhabited by the preexistent soul at the moment of conception,[74] and is the source of the passions (lust, anger, etc.) against which we all must struggle.

The anthropology of Tertullian (born approximately 155 CE; died after 220 CE) was much more bipartite than that of Irenaeus (and that of the Apostle Paul). For him, humans comprise only body and soul: entirely distinct from each other, and not a united whole (as the ancient Hebrew concept of *nephesh* represented).[75] He emphasized that the human soul is itself a kind of body, albeit an immaterial and immortal one, and is distinguished from the Spirit of God.[76] The immaterial soul is created together with the physical body at conception and is born unclean (because of Adam) until reborn in Christ.[77] It then grows and develops during life from infancy to old age—much like the physical body itself grows—and develops differently in each person depending on their life experiences. When the physical body dies, the soul's further growth is arrested until the Second Coming of Christ. "Tertullian is the first Christian author to assert in so direct a manner this developmental characteristic of the soul. It is nowhere as clear in Irenaeus."[78] He sees the inability of certain souls to accommodate certain spiritual truths like a baby tolerating only milk for a while before adult food, and even refers to "the puberty of the soul."[79] The physical body is the temporary house of the soul while alive on earth, and will be returned to the soul immediately prior to the final judgment.[80]

In the fourth century, Gregory of Nyssa answered the paradox of mortal, passible, and gendered humanity being described as the image of a genderless, immoral, and pure Nature by also appealing to the twofold

---

73. Benjamins, "Keeping Marriage," 97–99; also see McGuckin, "Anthropology," 14; Dillon, "Origen and Plotinus," 7–26; Ferguson, *Encyclopedia*, 48.

74. Young, *From Nicea to Chalcedon*, 226.

75. Steenberg, *God and Man*, 64–67, 75–76; Green, *Body, Soul*, 17.

76. Edwards, "Early Christianity," 46; Steenburg, *God and Man*, 55–103; Murphy, introduction to *Neuroscience and the Person*, iv.

77. Dunn, "Roman and North African Christianity," 156; Steenburg, *Of God and Man*, 68–72.

78. Steenburg, *Of God and Man*, 71–73.

79. Steenburg, *Of God and Man*, 73.

80. Dunn, "Roman and North African Christianity," 156; Steenburg, *Of God and Man*, 64–65, 68–71; Green, *Body, Soul*, 17.

creation of our own nature, "a compound of an intelligent part akin to the divine and an irrational part, related to our bodily form and divided into male and female, akin to animality."[81] His Stoic mind forced him to see the need for Adam and Eve to be covered with animal skins as evidence that these transcendent psychic beings had fallen to a degenerate level that was now concerned with earthly matters.[82]

Athanasius of Alexandria (fourth century) saw the soul's primary function being to control the body, guiding it toward the good or toward the bad, toward heaven or toward earth, in response to impulses it received from the *nous*.[83] In this way, he began to probe a question which would bewilder later thinkers such as Descartes in the seventeenth century and later substance dualists into the modern day:[84] exactly how does the immaterial mind exert an action upon the material body? (We will examine this question in much more detail below.) He emphasized the link between anthropology and Christology, and ever "afterwards this nexus of ideas became inextricable."[85] Humanity was completely recreated and renewed with the arrival of Christ, the incarnation of the divine *Logos*.[86] Patristic thought from this point onward was now focused upon the Christ-nature and the idea of christological regeneration.[87]

In the late fourth and early fifth centuries, three Christian theologians each made very specific and unrelated statements regarding the question of the human mind or soul. John Chrysostom dictated that human sexuality did not exist before the fall, and that "man lived the life of angels in paradise and this angelic life is nowadays restored by those who follow their vocation of virginity."[88] Cyril of Alexandria tied the taking of the Eucharist to the regeneration of the image of Christ within us.[89] Apollinaris of Laodicea claimed that Jesus could not have had a human mind; rather, Jesus had a human body and lower soul (the seat of the emotions) but a divine mind.

---

81. Benjamins, "Keeping Marriage," 99.

82. McGuckin, "Anthropology," 14.

83. Steenberg, *Of God and Man*, 170–71.

84. Murphy, introduction to *Neuroscience and the Person*, iii and xi.

85. McGuckin, "Anthropology," 14.

86. Athanasius, *On the Incarnation*, 3; McGuckin, "Anthropology," 14.

87. McGuckin, "Anthropology," 14; Shults, *Reforming Anthropological Theology*, 140–52.

88. Benjamins, "Keeping Marriage," 101.

89. McGuckin, "Anthropology," 14.

"After the time of Jerome (ca. 347–420) the soul was generally thought to be created at the time of conception."[90]

Nemesius of Emesa is credited at this time for a compendium—*On Human Nature*—describing "the fundamental constitution, purpose, faculties and potential of humanity, an exploration carried out by means of the then conventional scholarly methods and probably drawing all its material from standard textbooks of the time."[91] He is mentioned here not because he introduced something new to our understanding on this question of body and soul, but because he integrated roughly one thousand years of accumulated thinking on the subject. His overall conclusion is that humans are a psychosomatic whole—a fusion of body and soul which can be separated (as in death) but which are intimately united during life. The soul inhabits every part of the body, and motivates the body, but is not confined by the body: in fact, it remains a part of the universal mind. Because humans are in part material, we are mutable; because we are endowed with reason and can contemplate actions and consequences, we have free will. The latter allows us to tame the passions, and through contemplation of God, remain immutable (a Platonic goal). Sin is not inherent in our nature, but is a consequence of bad choices. Clearly, Platonism, Stoicism, and Neoplatonism are now interwoven in Christianity.

The young pre-Christian Augustine (354–430 CE) wrestled with the Manichean idea that the soul is "a spark of the Divine Light, of one substance with God . . . captured and carried off in a primordial invasion of the Divine by the hostile forces of evil. As the result of that primordial catastrophe, it found itself now in misery, immersed in a body which was Matter, hence darkness, evil."[92] After much deliberation over this and many other theological/philosophical matters, Augustine rejected Manichaeism and converted to Christianity in 387 CE, just a few years before Constantine declared Christianity to be the religion of the Roman Empire in 391. He then began a journey of using Platonic and Neoplatonic thinking to rebuild his theology. Navigating between the Platonic understanding of the relationship between body and soul (a strong dichotomy) and the Aristotelian view (the soul is the Form of the body, and therefore the two are inseparable), he settled on a Neoplatonic view (the soul is easily separated from the body) to help him explain/understand the fall of humans, as well as the resurrection of the soul and an "intermediate" state after death.[93] Augus-

---

90. Murphy, introduction to *Neuroscience and the Person*, iv.
91. Young and Teal, *Nicea to Chalcedon*, 222.
92. O'Connell, *St. Augustine's*, 88–93.
93. Shults, *Reforming Anthropological Theology*, 166–67; O'Connell, *St. Augustine's*,

tine "emphasiz[ed] the care and development of the soul as the means of salvation.... It is by cultivating the higher faculties of the soul (and often by repressing the lower faculties and the body) that one develops the capacity for knowledge of and relation to God."[94]

He wrestled for years to understand the origin of the soul, finding four possibilities. Two of these posit the soul coming into existence at the time that the body forms, either arising through generation (hearkening to Plato's Forms and Ideas) or alternatively being newly created when the new baby is conceived or born. The other two possibilities have the soul being already preexistent before the body, and either being assigned by God to a given fetal body, or willingly choosing to enter a given fetal body.[95] Although Augustine saw the body as good (being a creation of God), it was the soul's choosing to leave the spiritual and to join to the material which represented "the fall" (a betrayal) for which we are all guilty by virtue of the simple fact that we all made the choice to now exist with material bodies.[96]

Augustine also built on Plato's and Aristotle's differing models of a tripartite structure for the human soul. Whereas Plato divided the latter into rational, spirited, and appetitive aspects (*logistikon*, *thymikon*, and *epithymetikon*, respectively), and Aristotle divided between the human, animal, and vegetable souls (rational, sensitive, and nutritive, respectively), Augustine saw the three parts of the human soul being reason, memory, and will.[97] Furthermore, to him, they were "an image of the Trinity because they are one essentially, but three relatively ... are equally substantial with one another, and together (in their unity) they constitute the one substance of the human mind."[98] Augustine focused heavily upon christological anthropology: it became "a lens by which to illuminate every other problem of Christian thought."[99]

109–31; Turner, *Thomas Aquinas*, 72–73; Murphy, introduction to *Neuroscience and the Person*, iv–v.

94. Murphy, introduction to *Neuroscience and the Person*, v.
95. O'Connell, *St. Augustine's*, 149.
96. O'Connell, *St. Augustine's*, 146–83.
97. Schults, *Reforming Anthropological Theology*, 169–70; Hochschild, *Memory*, 12.
98. Shultz, *Reforming Anthropological Theology*, 170.
99. McGuckin, "Anthropology," 14; also see Benjamins, "Keeping Marriage," 103.

Human Nature from the Pre-enlightenment to Today

The prevailing influence on Christian thinking from the fifth to thirteenth centuries CE was Neoplatonism. The Greek-thinking fathers of the church had developed human ontology markedly from that which can be gleaned from biblical literature, and Augustine left his indelible stamp upon it. The millennium which followed saw further discussion on this subject by many notable thinkers from the sixth century and onward (Boethius, Nestorius, Eutyches, and Leontius of Byzantium to name just a few).[100] The latter thinkers did not produce entirely new directions or perspectives on the question, as occurred during the Hellenistic period or, as we will see below, during the age of modern science. Instead, they provided more finishing touches upon Patristic thinking in part by developing newer terminology more appropriate to their own unique cultures and ever more precise dissection of certain subtler points. An emphasis was placed upon the stark contrast between the material body/brain and the immaterial mind/soul. These were seen to be two essential components of the human being, but were entirely different substances, and so this view is commonly referred to as substance dualism.

In the thirteenth century, Thomas Aquinas (1225-1274 CE) reintroduced Aristotelian thinking,[101] which Augustine had rejected. In particular, Aquinas taught that the body is matter which has become temporarily activated by an eternal spiritual substance (a Form)—the soul—which shapes, controls, and animates the body while also providing consciousness and a motivation to pursue (or reject) goodness.[102] He wrote: "Body and soul are not two actually existing substances; rather, the two of them together constitute one actually existing substance."[103] Pasnau insists that "Aquinas is not a dualist, not even when dualism is understood along the lines of property dualism rather than substance dualism. Human beings are not the composite of two fundamentally different kinds of properties or entities. But of course Aquinas is not a materialist, either. He rejects materialism because he believes the rational soul is both incorporeal and subsistent."[104]

---

100. Shults, *Reforming Theological Anthropology*, 170–72.

101. Scott and Phinney, "Relating Body and Soul," 92–93; Pasnau, *Thomas Aquinas*; Murphy, introduction to *Neuroscience and the Person*, v; Anderson, "On Being Human," 184–85; Barbour, "Neuroscience," 252.

102. Turner, *Thomas Aquinas*; Pasnau, *Thomas Aquinas*, 65–72; Ward, *Big Questions*, 139; Murphy, introduction to *Neuroscience and the Person*, v; Peters, "Resurrection," 314; Moreland and Rae, *Body and Soul*, 202.

103. Aquinas, *Summa Contra Gentiles*, II:69:2.

104. Pasnau, *Thomas Aquinas*, 71.

Neither did Aquinas (or Aristotle) see the body being *united* with the soul, in the holistic sense that some Christians now believe, but rather is a vessel for the soul; to be more precise, the soul resides in the heart (while the brain is needed for cooling the blood).[105] Strangely, he believed "that the rational soul is infused by God into the body at 40 days for males and 90 days for females,"[106] and, in order to account for the fact that a soul was needed to form a body during the first few weeks of gestation, that there were successive "types" of souls: the vegetative/vegetable soul being involved in the earliest stages of conception and fetal development, while the sensitive soul further develops the fetus and prepares it for reception of the rational soul from the hands of God.

René Descartes (1596–1650) sought to move away from the tripartite model of the immaterial soul that Aristotle and Aquinas promoted, and instead move toward a more materialistic explanation. He now saw the brain as responsible for many of the faculties that had previously been attributed to the nutritive, sensitive, and animal souls. But lacking a fully complete and coherent model, he divided the human essence between the "thinking thing" (*res cogitans*) and the "extended thing" (*res extensa*), and promoted the former as the "real" person (the soul, which he saw as identical to the mind) (see figure 1).[107]

---

105. Pasnau, *Thomas Aquinas*, 37.

106. Scott and Phinney, "Relating Body and Soul," 92–93; Barbour, "Neuroscience," 252.

107. Shults, *Reforming Anthropological Theology*, 174; Scott and Phinney, "Relating Body and Soul," 93; Allen, "Persons," 165; Torrance, "What Is a Person?" 199; introduction to *Neuroscience and the Person*, x; Anderson, "On Being Human," 185, Peters, "Resurrection," 311; Barbour, "Neuroscience," 252–53.

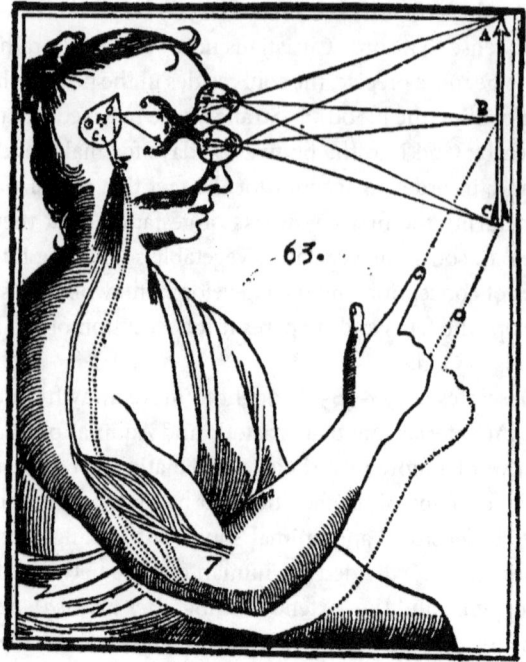

Figure 1

Reproduction of drawing made by René Descartes, showing an object (here an arrow marked by the letters A, B, and C) being perceived through the eyes of a person (the *res extensa*). The image of that object, projected onto the retina of the *res extensa*, is then observed by some kind of consciousness inside the head of that person (the *res cogitans*), which interprets the image and sends commands to the muscles of the arm and hand of the *res extensa* to point at the object. Image taken from the public domain.

Moreover, he strengthened the view of substance dualism: the mind/soul and the body are two distinct substances (immaterial and material, respectively), each of which can exist without the other but which together fully complement each other. In some ways, his view of the mind/soul in relation to the body is much like the farcical image presented in the 1997 Hollywood sci-fi action comedy *Men in Black*, in which the character Agent J played by Will Smith opens up the face of a dead body in a morgue to reveal inside a tiny, mega-cerebral Arquillian alien by the name of "Rosenberg," operating gears and levers and saying with his final dying breath that "the galaxy is on Orion's belt." Daniel Dennett, in *Consciousness Explained*, repeatedly refers to (and pillories) this view of the mind-body duality as the "Cartesian theater," in which the Ego is the "audience" and the sensory inputs and cognitive outputs are the actors on the stage of consciousness.[108] Although Descartes made many contributions to this question (and to many other fields of science), he is perhaps most popularly remembered for two in particular. The first is his often-quoted declaration which is the pride of proponents of substance dualism: *"cogito, ergo sum"* . . . "I think, therefore I am."

The second is also forever remembered, but more to their chagrin: in it, he addressed the vexing problem of the interaction between the material and immaterial. His world was one in which science was beginning to blossom and produce all kinds of technological wonders; a world of cause and effect explained by physical principles. As his peers focused their attention on cognition, they understood that the mind might be the impetus behind contraction of a muscle, but could not comprehend how the immaterial could exert any kind of material effect. They took note that the muscle was innervated by one or more neurons emanating from the spinal cord, which in turn emanated from the brain. From their growing understanding of hydraulics, they drew from the analogy of a church organ, one of the leading technologies of their time (much as we do today when we compare the brain to the leading technology of our own time: the computer). They imagined the muscles being actuated by some kind of hydraulic pressure transmitted by the brain via the innervation. A remaining question, however, was the location and nature of the "organist" playing the keys: Descartes traced this back to the pineal gland buried deep within the brain.[109] Not only was his hypothesis completely wrong, it did nothing to describe the interface between the material and immaterial (how could such a small organ exert enough pressure to move the arms and legs?).

The Reformation of the fifteenth/sixteenth centuries made relatively little impact on this question of the nature of the soul, in contrast to the tremendous impact it had on so many other areas of Christian theology.[110]

In the eighteenth and nineteenth centuries, New Testament scholars began to question the concept of resurrection, which in turn "led to increased emphasis among theologians on the immortality of the soul as the basis for Christian hope in an afterlife."[111] Nancey Murphy, Ray S. Anderson, and Joel B. Green separately provide brief but helpful historical summaries of the "confused" multiplicity of viewpoints which have been claimed to be

108. Dennett, *Consciousness Explained*, 253–55, 257.

109. Cortez, *Theological Anthropology*, 75; Shults, *Reforming Anthropological Theology*, 174; Ward, *Big Questions*, 143; Murphy, introduction to *Neuroscience and the Person*, xiii; Dennett, *Consciousness Explained*, 34.

110. Murphy, introduction to *Neuroscience and the Person*, v.

111. Murphy, introduction to *Neuroscience and the Person*, v.

widely accepted Christian dogma during this time and extending into the early twentieth century.[112] All three draw out a theme in which they perceive a gradual shift from dualist to monist thinking in the writings of scholars such as Karl Barth, Rudolf Bultmann, and John A. T. Robinson. However, this transition in thought has not completely won over the scholarly community—some modern theologians and philosophers continue to be strongly committed dualists, especially those of the evangelical tradition[113]—and it has barely penetrated the lay community: believers will often gather around the casket at the front of the church or at the graveside and point toward it saying words which express the view that "this is not our beloved sister/brother . . . they've gone on to be with the Lord, leaving this shell behind." Does this not reveal an acutely dualistic perspective, and a view that the body is unimportant? Likewise, there is a general conviction in the lay community that a core element of salvation is leaving the material confines of earth and gaining a spiritual existence in heaven—a very Platonic viewpoint—which scholars such as N. T. Wright are trying to counter.[114]

Modern theologians and their scholarship might indeed be exerting some influence upon Christian theology which has not yet fully matured, but their influence is being dwarfed by that of modern scientists.

Modern Science and the Church
Duel over the Immaterial

From the Age of Enlightenment to the present day, the scientific enterprise has worked to do away with Platonic/Neoplatonic/Cartesian thinking and Aristotelian/Thomist thinking, based upon Ideas, Forms, spirits, and the immaterial. Along with that, it seems to be displacing Christian thinking as well. More precisely, it has introduced yet another completely different paradigm which is still in the process of guiding or even coercing a complete revision of Christian thinking on the subject of anthropogony and human ontology. Where Aristotle (and in a sense Plato, as well) and Aquinas saw a nutritive/vegetative soul which controlled growth, nutrition, and reproduction, modern scientists locate the pituitary hormones and a variety of neurochemicals (for appetite and sex drive); where the former group saw an animal soul controlling movement and sensory perception, the latter group locate the

---

112. Murphy, introduction to *Neuroscience and the Person*, v–viii; Murphy, "Human Nature," 20–22; Anderson, "On Being Human," 185–87; Green, *Body, Soul*, 4–21.

113. Swinburne, *Evolution of the Soul*, 145; Eccles, *Brain and Conscious Experience*, 312, 327; Moreland and Rae, *Body and Soul*, 23, 39; also see Green, "Bodies," 151.

114. Wright, *Surprised by Hope*; Wright, *Day the Revolution*.

premotor and motor cortical regions and the cerebellum and several other brain regions; where the former saw the rational or human soul, the latter also point to the frontal cortex and various other brain regions.

The forward march of science displacing belief in the immaterial may be seemingly inexorable, but it has nonetheless also seen some interesting compromises, collaborations, and dead-ends. One example would be phrenology, the "science," by which it was claimed one could assess cognitive abilities such as intelligence or creativity by feeling the bumps on someone's head. A more relevant and thoroughly entertaining example might be an enterprising young physician-scientist practicing medicine in 1901 in Haverhill, Massachusetts, who set out to perform a provocative experiment: to weigh the human soul. More surprisingly, he obtained what he felt was a definitive answer—just over 20 grams—which was dutifully published in the mainstream and scientific press.[115] (Perhaps the ancient Egyptians were not so far off the mark with their belief that the soul of the dead needing to be weighed in the scale of Maat against the feather of truth before a decision was made about that individual's fate in the afterlife?)

Before the reader reacts too quickly—positively or negatively—to this anecdote, it is important to learn the details. Dr. Duncan MacDougall examined six patients who were suffering from diseases with symptoms which included physical exhaustion; he felt the latter condition was an important inclusion criterion for the study, since the patients needed to remain absolutely still while they died in order to get accurate measurements of weight. When death appeared imminent, the patient's bed was placed on one side of a massive balance—essentially a see-saw which he constructed—with an equivalent amount of weights on the other side of the balance to center the beam. As the patient lay there during the final hours of their life, any changes in weight were then measured by placing small counterweights on the opposite side of the balance: MacDougall claimed that he could in this way measure changes as small as two tenths of an ounce. He reported that death resulted in a loss of weight of roughly half an ounce in four patients: he discounted his measurements in the remaining two patients for technical reasons (the equipment was claimed to be "not finely adjusted" in one case, and there was physical interference by bystanders in the other). However, it is puzzling that he described how, in two of those four cases, the reported drop in weight occurred in two steps separated by as much as fifteen minutes. Even more concerning was what he reported for the third patient: the beam deflected upon death, as measured by the addition of the

---

115. *New York Times*, "Soul Has Weight"; MacDougall, "Hypothesis," 240–43.

small weights, but when those weights were later removed the beam did not deflect back again until fifteen minutes later.

Not surprisingly, MacDougall's claims were vigorously contested. Some attributed the weight changes to fluid loss through sweating and breathing; MacDougall defended against this. I myself have very little confidence in MacDougall's methodology and final conclusion. The magnitude of the reported changes (0.375 to 1.0 ounces, equivalent to the weight of a tablespoon of water) barely exceeded the self-acknowledged resolution limit (0.2 ounce) of the equipment that he cobbled together. Moreover, that magnitude of change was dwarfed by the overall weight of the patient and their hospital bed (this is akin to trying to measure the dimensions of a grain of pollen using a standard 12-inch ruler). MacDougall also draws particular attention to the suddenness of the phenomenon, referring in his scientific paper to the audible stroke of the scale's beam hitting the lower limiting bar, and in his interview with the *New York Times* saying, "The instant life ceased, the opposite scale pan fell with a suddenness that was astonishing—as if something had been suddenly lifted from the body." Again, however, the relative numbers just call this into question: the kinetic energy of a sudden loss of half an ounce (or 20 grams) accelerating the much larger mass of the patient and their hospital bed (100 kilograms, or 100,000 grams) to such a speed that it makes an audible sound is hard to believe: it is actually less than the acceleration that would be seen in a billiard ball (160 grams) being impacted by an adult fruit fly (0.22 grams).

These facts, together with the observation that the small deflections could take up to fifteen minutes to register (or even not revert back upon removal of the counterweights), lead me to think instead that these deflections were more related to an accumulation of very small amounts of tension within the levers and gears of MacDougall's apparatus as the patient lay there for hours—shifting body positions, coughing, slowly losing fluid by breathing or coughing, and being very gently manipulated by MacDougall listening/feeling their heartbeat (he reports that he did so in the scientific publication)—and all these small accumulations of tension would from time to time exceed the internal friction of the apparatus and be released (the beam would deflect ever so slightly when that friction gave way). Tellingly, MacDougall did not increase his sample number beyond the four that he included in his results, even though he himself admitted that this was necessary before a more firm conclusion could be made, nor did anyone attempt to repeat and corroborate the study. However, that has not stopped vigorous attempts to defend MacDougall's findings scientifically.[116]

---

116. Ishida, *Rebuttal*.

In the next chapter, we will explore several lines of thinking which attempt to explain human consciousness, mind, and soul entirely through naturalistic mechanisms: ones which arise solely from the constituents and properties of the neurons in our brain. We will see how all of the cognitive faculties which were previously attributed to our soul or spirit (such as will, emotion, memory, and sense of self) are housed within specific regions of our brains; moreover, that the activities of those brain regions can be modulated by specific chemicals (neurotransmitters, such as serotonin), some of which can artificially induce a profound "spiritual" experience (hallucinogens). Moving further into molecular mechanisms, we will explore a novel hypothesis that consciousness may be a product of subtle physical shifts and energy resonances in a structural molecule known as tubulin, modulated by quantum mechanical effects (entanglement; superposition). Finally, we will see that genetics can play a role in our spirituality.

It is now widely accepted in today's world that the material aspect of our being—our physical body—arose through the process of biological evolution over billions of years (strictly speaking, this contradicts an instantaneous sculpting of clay or mud, but an open-minded person could acknowledge that our being sculpted from dirt is indeed a metaphorically or mythically accurate statement: as Carl Sagan once said, "The cosmos is within us. We are made of star-stuff. We are a way for the universe to know itself."). Many diverse lines of evidence attest to this theory: space constraints prevent a summary of them here (though I have done so in a previous book)[117] and the reader can consult many other sources.[118]

Pew Research Center have regularly conducted polls which, among many other things, probe the values and belief systems of different people groups and track changes in the same over time. The most recent iteration of their polling venture was done in 2014, although an earlier version was done in 2009, and yet others at various periods over the past many decades. Interesting comparisons were made between groups and time periods, exposing trends in public and Christian thinking. In this most recent case, Pew talked to 2,002 members of the general public, as well as to 3,748 members of the premiere scientific organization in the United States: the American Association for the Advancement of Science (AAAS; includes nearly 250 affiliated societies and academies of science, serving ten million individuals worldwide).[119] It is important to emphasize that roughly half of the AAAS

---

117. Janssen, *Standing on Shoulders*.

118. Harlow, "After Adam," 179–95; Stearley, "Assessing Evidences," 152–74; Venema, "Genesis and the Genome," 166–78; Wilcox, "Proposed Model," 2–43; Finlay, "Human Evolution," 103–14.

119. American Association for the Advancement of Science, "About AAAS."

group include people who have extensive scientific training but do not actually do scientific research: they may work in non-research capacities in drug companies, in various government agencies, or as science teachers at high schools and colleges, for example. Even more importantly, it should be kept in mind that a significant fraction of this AAAS group comprises religiously minded people, including Christians.[120]

One component of this Pew poll was an open-ended question: those who felt that science does indeed conflict with their personal religious beliefs were asked to specify what particular beliefs, opinions, or issues were impacted. As might be expected, a very wide range of answers were given.[121] A large proportion of these (36%) named issues which pertain directly to the first chapter of Genesis: big bang cosmology and the evolution of life forms. Interestingly, and perhaps more disturbingly, a quarter of the respondents (18% plus 8%) could not (or would not?) clearly articulate specifically what, how, or why. Only 1% of the respondents pointed to "belief in the afterlife," and none seem to have indicated anything pertaining to mind, consciousness, soul, or spirit.

Pew also asked the American public and the scientifically trained AAAS group twenty-two specific questions pertaining to science-related issues which can be divided into seven categories: climate and energy; public funding of science research; human evolution; biomedical issues; food; and space exploration.[122] At the same time, they collected a wide range of demographic details about the respondents: political ideology or affiliation; age; race or ethnicity; gender; level of scientific training (whether they had a scientific degree[s]); and religion. In the latter case, they asked about religious affiliation (Protestant; Catholic; Jewish; Muslim; Buddhist; other) as well as their level of attendance at their chosen place of worship (weekly; monthly; a few times per year; never) in order to gauge their level of commitment to that particular religious affiliation. Statistical analytical techniques were then used to assess the degree to which each of the demographic factors listed above influenced one's opinion on the particular scientific issue being probed. Only two of these twenty-two questions were strongly influenced by one's religious affiliation: human evolution, and global overpopulation.

Pew Research found that AAAS members agree nearly unanimously (98%) that humans evolved from lower hominids (90% believe this involved natural processes, while only 8% felt this was guided along by a supreme

---

120. Maskie, "Scientists and Belief"; Pew, "Elaboration of AAAS Scientists' Views."

121. Pew, "Perception of Conflict."

122. Funk and Masci, "5 Facts"; Funk and Alper, "Religion and Science"; Funk and Rainie, "Americans."

being, even though 51% indicated that they *do* believe in either God [33%] or in a higher power [18%]).[123] Among those AAAS members who are actively engaged in publicly funded scientific research (1,627 members; still approximately half of the AAAS group), that proportion rose to nearly complete unanimity (99%).

The general public, however, were less convinced: roughly a third of them (31%) believed that humans have always existed in their present form.[124] Further analysis showed that most of this disconnect between the general public and the AAAS group owed to religious affiliation: 36% of religiously affiliated individuals denied human evolution, and only half accepted it (26% believe it involved natural processes only, while 27% thought it was guided by a Supreme Being). When the specific details of one's religious affiliation were taken into account, Pew found that the greatest disconnect with the AAAS group was due to the contrary views held by Protestants, and particularly white evangelical Protestants: only 33% of the latter believed that humans have evolved, compared to 65% of white mainline Protestants and 62% of white Catholics. Black and Hispanic groups closely paralleled these differences.

Gallup Inc. have also conducted their own version of this kind of polling study, and overall made similar observations.[125] In particular, they found that nearly half of the American general public have consistently believed that humans were created by God in their present form (this value has hovered consistently between 40–47% during the period spanning 1982 to 2014), while most of the remainder have believed that humans evolved under the guidance of God (31–40% during that same period). Interestingly, there has been a steady rise in the percentage of Americans who believe that humans evolved by purely naturalistic mechanisms without any intervention on the part of God. The belief in humans having evolved was strongly influenced not only by age, but also by postsecondary education: the "creationist" view dropped from 57% among those with only a high school education to 27% of those with a college/university degree.

Another polling agency has shed additional light on the reasons for this resistance against the idea that God used the process of biological evolution: the Barna Group is a private, nonpartisan, for-profit organization "widely considered to be a leading research organization focused on the intersection of faith and culture," and has "tracked the role of faith in America, developing

---

123. Maskie, "Scientists and Belief"; Pew, "Elaboration of AAAS Scientists' Views."

124. Pew, "Elaboration of AAAS Scientists' Views"; Pew, "Evolution and Perceptions of Scientific Consensus."

125. Newport, "42% Believe."

one of the nation's most comprehensive databases of spiritual indicators."[126] In 2012, they polled 743 pastors of churches "big and small and from all Christian denominations" across the United States about their view on whether God used evolution. They found that over half of those pastors had "major concerns," primarily because it "undermines the authority of Scripture" (64%), "views portions of the Bible as non-literal, like Genesis" (62%), "raises doubts about a historical Adam and Eve" (61%), and "raises questions about how and when death and sin entered the world" (59%).[127]

Pew also found that *non*-scientists look into the minds of *scientists* and somehow perceive that those scientists are themselves deeply divided on several scientific issues. Even though the members of the AAAS are in essentially complete unanimity about humans having evolved, a third of the public instead believe that scientists are divided in opinion on whether humans evolved, and half of the public think scientists are divided over the big bang.[128] Further statistical analysis of the polling data showed that the bulk of the disconnect between what AAAS members think and what the general public think the AAAS group think is due to a misperception held by those with some kind of religious affiliation. The group who exert the biggest influence behind this misperception are again Protestants (36% believe scientists are divided on this), and again more particularly white evangelical Protestants (49% of these believe scientists are divided). This inaccurate perception is puzzling, given that in actual fact scientists are 99% in agreement that humans evolved over time. One possible explanation is that believers, when faced with the evidence for human evolution and still wanting to deny it, might be trying to resolve their own inner cognitive dissonance or reassure their readers and listeners by saying, "Hey, even scientists can't agree on this stuff!"

This is by no means the first time that the church has perceived a theological threat in the discoveries coming from the scientific enterprise and attempted to push back. Perhaps the most infamous precedent is that of the Copernican Revolution in the sixteenth century. The church had long promoted a cosmological model featuring an absolutely immoveable earth built on pillars which also supported a large overturned bowl over our heads. The sun, moon, stars, and birds moved across the face of that bowl, and "the waters above" were held back behind it; rain was the result of small holes or windows opening up in that bowl. This cosmic model of a three-layered earth (land overlying the "waters below," then hell) and three-layered

---

126. Barna, "What Is Barna?"
127. Biologos, "Survey of Clergy," highlight #5.
128. Funk and Alper, "Religion and Science," 3.

heavens (cosmic lights in front of the bowl holding back the "waters above," and then heaven) was based upon many passages in the Bible, believed to be the attestation of God himself on the matter. At the heart of the dispute was that the Bible clearly described the earth as being motionless and immoveable at the center of the cosmos, with all the other celestial bodies in motion around it. I have documented at length several theological and biblical reasons for holding this view, as well as the main reasons that the church felt it necessary to vigorously reject the scientific cosmological model proposed by Copernicus, Galileo, and many other scientists standing beside a mountain of compelling data.[129] Those reasons include:

- "It contradicts Scripture."
- "It goes against church tradition. We've always held the old way to be true: our leaders, called by God himself to shepherd us, have always taught us these things."
- "God would not work that way."
- "It means we are not unique among all created things."
- "It challenges the idea that we are created in the very image of God."
- "On the sixth day, God declared everything 'very good.' How can such a distortion of that which is right be very good?"
- "Jesus and the apostles clearly believed in the accuracy and historicity of Genesis."

However, Copernicus, Galileo, and many other astronomers proposed a very different model which more simply explained the complex motions of the heavenly bodies: one which featured earth spinning on its axis at a thousand miles per hour (as measured at the equator), and orbiting the sun at sixty-seven thousand miles per hour, caught in the gravitational well of the sun. Anything but motionless and immoveable. Eventually, the evidence for the scientific model grew too great to resist, and the church capitulated. It is easy for us to dismiss that part of our history; to smile or even laugh at it. But we should try to also learn from it. Goheen and Bartholomew drew the following conclusion:

> The first great mistake of the church was to react so negatively to the rise of the new science. . . . The Christian church could have responded differently. It could have asked if its traditional interpretations were correct; it could have rearticulated the Christian

129. Janssen, *Standing on Shoulders*, 36–38.

faith for a new time. . . . It did not have to be this way; religion and science are not in irreconcilable conflict.[130]

I would argue that the church is now repeating all these very same mistakes vis-à-vis the matter of human evolution, and for the very same seven reasons listed above. I would also argue that this is all quite unnecessary. Many have found it possible to maintain a Christian faith which embraces those new scientific discoveries regarding evolution. To do so, though, does require a new reading of Scripture and reinterpretation of many strands of Christian theology. The majority of clergy surveyed by the Barna Group agreed that, "just as scripture should influence human interpretation of science, science should also inform our understanding of scripture."[131] Christian thinkers now have the opportunity to move far beyond the very outdated conflict over "young earth" versus "old earth" worldview, as well as the fast-becoming-outdated conflict over biological evolution, and to shepherd in a Christian theology which is compatible with the theory of human evolution. But another closely related paradigm shift is on the horizon which even these forward-thinking scholars have not yet fully anticipated: having addressed the theological ramifications of embracing the scientific evidence pertaining to the *biological* aspect of human ontology—our physical bodies—what are we to do with the rapidly accumulating scientific evidence pertaining to our non-biological *inner* being?

---

130. Goheen and Bartholomew, *Living at the Crossroads*, 89.
131. Biologos, "Survey of Clergy," highlight #4.

CHAPTER 3

A Synthesis of Modern Thinking on Our Inner Being

Opening Matters: Terminology

Before making our way through this chapter, it will be useful to bring into crisp focus certain terms which are central to the discussion. These terms are sometimes misused, misunderstood, or equated with other terms which are not precisely equivalent.

Material versus Immaterial

The first of these two terms is *not* simply "those things which are composed of matter," at least not in the concrete way that most people understand matter. Einstein's famous equation—$E=mc^2$ (where E, m, and c are energy, mass and the speed of light, respectively)—informs us that energy itself is a form of matter, and that the two are easily interconverted and directly related (in much the same way that heat is a form of cold, or a financial credit is a form of debit). Material things are also those which can be studied, described, and explained using material means such as the tools of science. As such, forces like magnetism and gravity are material things even though they are invisible and cannot be held by a pair of tweezers or put into a test-tube. In contrast, immaterial things are not composed of matter, do not exert forces, and are immune to scientific investigation and explanation: such things include virtue, or the concept of the number eleven (or any number). That much may be reasonably clear to the reader.

But what of things which seem to be immaterial, but arise through material mechanisms? For example, consider cyberspace, or a virtual reality game. Both are completely explained by hardware (computer boards;

memory chips; logic circuits; wires; servers) and software (computer programs). But does this alone mean they are material? Cyberspace is not an actual place. As we fill cyberspace with pictures, videos, text, or data, it does not get any heavier, nor exert any greater amount of force of any kind. If the internet were to ever collapse completely, the weight of earth would not be any less. Neither can cyberspace be atomized: even though I access it using my laptop, I cannot break off a piece of cyberspace and carry it around in my laptop. Likewise, virtual reality games. These begin with a computer, which is completely material. But then you need software, which is simply information (immaterial) stored on something physical like a USB stick (material) or carried by electrons (material) downloaded from some website on the internet. You might be pausing here at my comment that the information is immaterial. Certainly the electrons and the magnetism are material, and they exert material changes to the computer's memory chips, but the *information* itself is *not* material. That information does not exist in the USB or the data streaming from the internet. A USB stick which is completely full of files with hugely important information is not measurably different in any conceivable way from one which is filled with files of gibberish ... until you plug it into a computer and it is read and interpreted by other chips in your computer and then realized within your brain. The computer builds up a pattern (immaterial) which can be read by other computer chips (material), which then build up arrays of information (immaterial) in RAM memory (material). Each individual array of information does not have a dimension of time, nor of spatiality (up-down; left-right).

The computer takes those non-dimensional immaterial arrays of information and converts them into a signal having dimensionality described by electrical charge and time (since it changes from microsecond to microsecond), but not innately having dimensions of left/right, up/down, or back/front. Chips controlling the monitor screen take that signal and use it to modify the light output of its screen, which is a material event (because of the electrons being converted to photons). But the light output of the monitor (which is material) happens to correspond to a two-dimensional image which is immaterial ... the brightness and color *do* exist in the absolute/real sense, but the *image* does *not* exist anywhere except in a brain that is capable of interpreting the light signal. An image of a magnet would not exert any force whatsoever on a piece of metal. An image of a nuclear reactor melting down would not cause any tissue damage.

That material light output with two spatial dimensions and one temporal one is received by your retina and converted into changes in the electrical state of your neurons which then modify the distributions of electrical charges in your visual cortex. All of this is purely material ... material substances (retina and neurons) interacting with material substances (light and electricity) using material substances (neurotransmitters) and material forces (electrical and chemical "batteries"). But those changes in the distributions of electrical charges within the visual cortex (material) are perceived by the visual cortex as a four-dimensional experience. That experience is purely immaterial: someone having a profound, mind-blowing experience is no more electrical nor heavier nor magnetic than someone who is bored to tears in a sensory isolation tank. One can still measure much of the same electrical changes and metabolic activity in the visual pathways of someone who is blind because of a neuroanatomical defect which cuts off those pathways from the parts of your brain which allow you to *experience* them. A child and an adult can look at the exact same two-dimensional light output from the computer monitor (material) but have a completely different four-dimensional experience (immaterial).

All these points apply equally when we consider emotion. Hatred is an emotion experienced in and by the human brain (it is unclear if other higher animals experience hatred), and therefore would not exist without that material organ, but someone who becomes increasingly hateful does not become heavier, nor do their brains exert any more force of any kind. Someone who is hateful (or loving) can themselves exert force, and an organization that is motivated by hatred (or love) can exert force, but the emotion which motivates the individual or the organization does not exert any force. The electrophysiological changes which occur in the neurons of the brain that is experiencing hatred are completely material, but the emotion being experienced is completely immaterial.

This concept of "immaterial" may be foreign and confusing to some readers. Nonetheless, I hope this extended set of paragraphs on defining it at the very least clarify that immaterial does not mean invisible and untouchable. The internet is material but cyberspace is not. Computers and human brains are material things, but information and emotions are not. The human spirit may be immaterial, but the human soul and mind do not have to be: below, we will explore several completely materialist or physicalist explanations for mind and soul.

The Inner Self

The reader will benefit from paying careful attention to the fact that the being one calls "myself" or "me" is not some monolithic thing. We will see below that mind, personality, consciousness, soul, and spirit are quite different concepts all wrapped up together in that term "self." Many who study cognitive science, being non-theists, may prefer to avoid speaking of soul or spirit, but are quite comfortable trying to understand and explain mind, personality, and consciousness. They would see those latter three concepts to be simply emergent properties of a highly developed and amazingly complex biological organ, the brain (we will explain the concept of emergence in the next section).[1] In this chapter, we will bring together the current thinking on these various aspects of the inner self and see that it is indeed possible to extend this idea of emergence to the concept of the human soul as well.

Monism versus Holism (or "Wholism")

Monism in the context of this essay points to humans being solely material in composition, in contrast to dualism which claims we are two substances: a mixture of material and immaterial. Asserting that humans operate as a psychosomatic unity is not necessarily equivalent to monism. This is an important distinction, as some seem to confuse the two. For example: "The [Old Testament] clearly presents the human person as a psychosomatic unity, and does not distinguish between material body and immaterial 'soul.'"[2] A functional unity can be comprised of more than one distinct entity or substance. For example, the "moss" which grows on ancient rocks—forming beautiful, colorful patterns on the surfaces of the boulders and cliff faces—is a symbiotic union of two entirely different kinds of living organisms: fungus and algae. Those two types of organisms are completely different from each other in so many respects, and originate from two completely different parts of the evolutionary tree; but when they join together in the form of a lichen, they take on yet other properties which are completely different from either organism alone. It is not that the one is "infected" by the other: they form a compositional and functional unity. Another analogy may be helpful: a professional tennis star may "feel" and operate as if the racket is an extension of their own body, and an elite jet fighter pilot may fly as if she and the airplane were "one"—a functional unity—but at the end of the day they will both discard those mechanical objects as not part of themselves and those machines will become completely separate, inanimate, and nonhuman. All three analogies present two different material components functioning together as one unity, but the principle underlying my use of them can apply equally to this question of monism versus dualism: a dualist view of ontology can still explain a functional unity of categorically dissimilar parts (material and immaterial).

1. Shults, *Reforming Anthropological Theology*, 179–80.
2. Johnston, "Humanity," 565; also see Green, *Body, Soul*, 10.

The Image of God, or *Imago Dei*

In discussions about our inner self, some people may get emotionally charged because of a certain attachment to the concept of humans being created in the image of God, and a perceived threat to that concept when science is brought to bear on the discussion. The book of Genesis introduces readers to the *imago Dei* (Gen 1:27), and it is used several times in the Old Testament and then picked up much later by the Apostle Paul (1 Cor 11:7; Ro 8:29; 2 Cor 3:18; Eph 4:24; 1 Cor 15:49) and the author of Hebrews (Heb 1:3). However, the meaning of the *imago Dei* is never clearly defined within the Bible. We have seen how (Greek) theologians from the Patristic era turned this into something quite different from what the (Hebrew) authors of Scripture seem to have intended: the Hebrews found it to pertain to Divine ambassadorial responsibility and a corporate authority of humans (male and female together) made manifest through relationships (with God, with each other, and with creation), while the Greeks individualized it and related it more to cognitive abilities (some even limited it to only males). Today, many Christians have extended the Greek understanding of the *imago Dei* even further, defining it as those cognitive aspects which they feel set humans apart from the animals: free will, volition, morality, reason, creativity, language, symbolism, empathy, love, and an artistic sense. They may be surprised to learn that many of these cognitive abilities are increasingly being identified in various animal species. We will see below that these can instead be seen more like cognitive "tools" which helped humans achieve relatedness to God, rather than as "markers" of our God-likeness. Otherwise, by elevating these cognitive features too much, we may risk creating God in our own image. Xenophanes, a sixth-century BCE Greek philosopher, said: "Thracians think that their gods are blue-eyed and red-haired, as they themselves are; the Ethiopians, by contrast, make their gods dark-skinned with broad noses, as, to be sure, they themselves are . . . if horses and oxen could draw, they would no doubt draw gods after their own self-conceptions, and those gods would, unsurprisingly, look like horses and oxen."[3] Many have already written at length on the *imago Dei* and how we can find the full and perfect expression of humanity within Christ:[4] space constraints prevent a recapitulation of that discussion here.

---

3. Shields, "Philosophy before Socrates," 6; Burkert, *Greek Religion*, 308–9.

4. Cortez, *Theological Anthropology*, 14–40; Middleton, *Liberating Image*; Steenberg, *God and Man*; Walton, *Lost World of Genesis One*; Wilcox, "Proposed Model," 2–43; Jeeves, "Neuroscience," 170–86; Garte, "Evolution and Imago Dei," 242–44; Berry, "Natural Evil," 87–98; Stearley, "Assessing evidences," 152–74.

The "Biblical" View: Monist or Dualist?

It is claimed that as many as 130 different views of human ontology have been proposed.[5] Unfortunately, while the Bible may rule out most of them, it just does not seem to be concerned with articulating a single, unified, theoretically rigorous theory or model.[6] Many scholars conclude that the Bible has always promoted a holistic view of humanity which is somehow different from a purely dualistic one: a unity rather than monism.[7] Here are three excerpts to illustrate how dogmatic they can be.

> Genesis denies that humans are bits of divinity stuffed into bodies as, say, some ancient Greeks thought.[8]

> The general consensus is that the biblical view of the human person is holistic, not dichotomistic as in Greek and some Hindu thought. This is not a simplistic opposition between Hebrew "monism" and Greek "dualism," but a recognition that there is in fact a difference.[9]

> The [Old Testament] clearly presents the human person as a psychosomatic unity, and does not distinguish between material body and immaterial "soul." ... [The traditional Christian view of] the human person as dichotomous ... has now been largely abandoned.[10]

However, can one claim so definitively that the Bible does not support a dualist view? Close scrutiny reveals that the situation is not so clear-cut. In order to address this question, we can track three different lines of evidence as we proceed through the Bible: (1) creation accounts which describe the constructing of humans; (2) passages which mention what happens to the

---

5. Cortez, *Theological Anthropology*, 69; Murphy, introduction to *Neuroscience and the Person*, v.

6. McGuckin, "Anthropology," 13; Scott and Phinney, "Relating Body and Soul," 91–92, 100.

7. Preuss, *Old Testament Theology*, 110; Childs, *Old Testament Theology*, 199; Cooper, *Body, Soul*, xv; Berger, *Identity and Experience*, 6; Cortez, *Theological Anthropology*, 21, 70; Shults, *Reforming Anthropological Theology*, 175–77; Clouser, "Reading Genesis," 246–48; Barbour, *Religion and Science*, 270–72; Green, "Restoring the Human Person," 1–22; Green, "Bodies," 149–73; Jeeves, "Neuroscience," 170–86; Siemens, "Neuroscience," 188; Scott and Phinney, "Relating Body and Soul," 100; Murphy, "Human Nature," 2–3; Jeeves, "Brain, Mind, Behavior," 91; Anderson, "On Being Human," 175–94; Barbour, "Neuroscience," 250; Green, "Body, Soul," 7–8.

8. Clouser, "Reading Genesis," 246.

9. Shults, *Reforming Anthropological Theology*, 175.

10. Johnston, "Humanity," 565.

inner being at the moment the body dies; (3) passages which describe resurrection and/or some kind of afterlife existence which can be experienced with or without a physical body.

The Torah

The first five books of the Old Testament constitute the literary and historical foundation of the Old Testament, and reflect aspects of ancient Hebrew thinking from a time before Israel became a nation. The second chapter of Genesis provides three important glimpses into this question of human ontology. In its description of the creation of the first human, we find a very dualistic representation: God first formed a human body from the dust of the ground, then breathed the breath of life into the man's nostrils (Gen 2:7). This passage then concludes with the life-giving breath and creative force transforming the newly created human into a spiritual being: the lump of clay was not *given* a living spirit, but the created man *became* a living spirit (Gen 2:7). Unfortunately, this glimpse is equivocal: it describes two distinct parts (a dualist image) which were brought together in a way that became a single thing which was quite different (certainly a holistic [and dualist] view which moves one's thinking toward monism).

A few verses later yet, when he creates the first woman, we are told that God simply took "one of the man's ribs" and then made a woman from it. There is no mention of an infusion of the breath of life nor any other component which might push us toward a dualist view. It seems instead that everything needed to make a human was found within that rib itself: a monist view.

A fourth clue is given in the third chapter of Genesis, describing the fall of mankind. God gives the injunction to not eat the fruit of a certain tree in the middle of the garden, and informs Adam of the outcome of breaking that single command:[11] Adam is simply told, "You will die . . . you [will] return to the ground, since from it you were taken; for dust you are and to dust you will return" (Gen 3:3, 19). There is no mention here of an immaterial component (the soul and/or spirit) leaving the decomposing body and returning to some kind of existence elsewhere; no transformation into some new kind of body; no judgment scenario, nor residence in some kind of place which is described as heavenly or hellish; no conscious after-death existence of any kind. Just simply that the "you" which was Adam would "return to the ground." This is also strongly monist language.

---

11. Some interpret this as a "penalty" for breaking that Divine command, but it is worded as merely a consequence or outcome rather than a punishment; see page 43.

However, a dualist view begins to reemerge as we progress through the remainder of the Torah. Several men are said to have been "gathered to his people" or "slept with his fathers" (this is never said of any woman) when they died and their bodies had decomposed, including Jacob (Gen 25:7–8; 35:28–29), Ishmael (Gen 25:17), Moses (Num 27:13; 31:2; Deut 32:50), and Aaron (Num 20:24; 27:13; Deut 32:50). These passages provide the first hints of an afterlife existence in Sheol (we will explore this in much greater detail in chapter 4). Furthermore, at least one person—Enoch—seemed to have bypassed death entirely: he was simply "taken away" and "was no more," and this is presented as a reward for having lived an exemplary life (Gen 5:24).

As time progressed, it seemed that the early Israelites began to pick up conceptions and rituals from the surrounding ancient Near Eastern nations and thereby developed their own death cult tradition (see page 133) which required the development of Deuteronomic and Holiness Law Codes to control it (Exod 22:18; Lev 19:31; 20:6, 27; Num 25:2; Deut 18:10–11; 26:14).[12] In other words, they came to believe that there were disembodied beings with very distinct personalities: the latter seemed to share a common history with the living family members, and therefore identified with and sympathized specifically with them.

It is clear, then, that the Torah begins with a confusing mixture of monist and dualist views of human ontology, but eventually dualism becomes somewhat more prominent.

The Former Prophets

The second major section of the Hebrew Bible is referred to as "the Prophets," which in turn is divided into three parts—the Former Prophets, the Latter Prophets, and the Twelve, or Minor, Prophets. The books of the Former Prophets—Joshua, Judges, Samuel, and Kings—constitute a continuous narrative of the period in Jewish history in which the nation of Israel occupied the land promised by YHWH. During this period, there did not seem to be a marked change in the development of Hebrew understanding of human ontology beyond that already revealed in the Torah. That is, in these books we also find numerous references to: death of the material body being the result of being deprived of a seemingly immaterial breath of life (Josh 10:40; 11:11; 1 Kgs 15:29); the essence of the person continuing beyond death in the form of "sleeping with his fathers" or "being gathered to his people" (Judg 8:32; 1 Kgs 2:10; 11:43; 2 Kgs 8:24; 22:20; 24:6; 1 Chr 29:28; 2 Chr 27:9; 34:28); a postmortem existence of some kind in Sheol (1 Sam 2:6;

---

12. Johnston, *Shades of Sheol*, 152–53, 169–70.

2 Sam 12:23; 14:14); and at least one person (Elijah) bypassing death and going "up to heaven in a whirlwind" (2 Kgs 2:11). However, the vague dualism of the Torah gains certain added dimensions in these books. We find here the story of Elijah raising to life the son of the widow at Zarephath, saying that the boy's *nephesh* returned (1 Kgs 17:21-22). There are also curious references to the spirit of one person, or a portion thereof, being transferred to someone else (Num 11:17-25; 27:20; 2 Kgs 2:9-15), or being transported far away from the body in dreams and visions (2 Kgs 5:26).

The Latter Prophets and the Twelve

The Latter Prophets (Isaiah, Jeremiah, Ezekiel), and the Twelve (Hosea, Joel, Amos, Obadiah, Jonah, Micah, Nahum, Habakkuk, Zephaniah, Haggai, Zecharia, Malachi), on the other hand, point to a period when the nation of Israel was being expelled from her land and entering into captivity. Within these books, a dualist perspective strengthens further. We find here, for example, increasing references to a heterogeneity of composition or multiplicity of parts vis-à-vis human ontology. Ezekiel refers to his spirit being transported far away from the body in a vision (Ezek 8:3). Isaiah refers to the wasting away of both *basar* and *nephesh* (Isa 10:18), and says it is the spirit of the man which seeks God, as well as his soul and heart (Isa 26:9). Zechariah tells us that "God forms the spirit within a person" (Zech 12:1).

We also see additional references to prayers being made to/for the dead (Isa 8:19; 57:5-7, 9; Ezek 43:7, 9), a much greater awareness of a conscious existence in Sheol (Isa 14:9-10, 26:14; 38:18; 59:10; Ezek 32:22-30; Jer 51:39, 57; Jonah 2:5), and references to the Valley of Gê-hinnôm or Gehenna (Isa 31:9; 39:31-33; 66:24; Jer 7:32; 19:5-7), which would centuries later develop into the place we now call hell (and which some would describe as a place of eternal conscious torment).

We will explore in chapter 4 the concept of Sheol being a common destiny for both the wicked and the righteous and how this would catalyze the development of the ideas of postmortem resurrection and judgment. There are hints in these books of this development beginning to stir, although the context of the passages leads scholars to suggest that these refer to the rebirth and renewal of the nation of Israel rather than resurrection of individuals to an eternal existence, and there is dispute as to whether it should be read literally or metaphorically. These passages include Ezekiel's vision of the valley of dry bones (Ezek 37) and the passage which precedes this ("I will give you a new heart and put a new spirit in you; I will remove from you your heart of stone and give you a heart of flesh. And I will put my

Spirit in you"; Ezek 36:26–27), and a similar one in Isaiah ("But your dead will live, Lord; their bodies will rise—let those who dwell in the dust wake up and shout for joy"; Isa 26:19).

Altogether, the Latter Prophets clearly present a dualistic perspective: humans comprise a material body which dies and decomposes, but their essential being is found in an immaterial component/substance that continues on in Sheol. Furthermore, there is a growing sense that the personhood would eventually be judged and rewarded for deeds done in the long-gone body.

The Writings

The third major section of the Hebrew Bible is referred to as "the Writings," and include the remainder of the books of the Old Testament. Here a decidedly dualist view surges. The psalmist clearly distinguishes between the spirit, body, and mind: "their *spirit* departs, *they* return to the ground; on that very day *their plans* come to nothing" (Ps 146:4; italics added in order to emphasize the reference to spirit, body, and mind/soul, respectively). Likewise, the author of Ecclesiastes writes that, upon death, "the dust returns to the ground it came from, and the spirit [*ruach*] returns to God who gave it" (Eccl 12:7). In the biblical story of the testing of Job, the Lord (Adonai) and Satan (the Accuser) distinguish between Job's flesh and his soul (Job 2:4–6), and the writer of a related apocryphal book indicates clearly that Satan was given authority over Job's body but not over his soul (*Testament of Job* 20:3). We also find other references to the spirit leaving one's body in order to receive a divine revelation, and then returning to the body (Ezek 8:3; Dan 8:2).

David and Job refer to their *nephesh* and their *basar* experiencing different, albeit complementary, feelings (Ps 63:2; Job 14:22). David and Ezekiel seem to differentiate between the *lebh* and the *ruach* (Ps 51:10; Ezek 11:19; 36:26); in Job, there is a clear distinction between Job's flesh and his soul (Job 2:4–6). A creation passage in the Psalms refers to mankind being made a little lower than the angels (who do not have a physical body) and yet above the animals (which presumably do not have a soul) (Ps 8:5–8). These books abound with references to a conscious existence in Sheol; more importantly, though, we find here testimonies of those who claim to have in fact been brought back from Sheol (Pss 18:4–16; 30:3; 40:2; 49:15; 71:20; 86:13; 107:10–22).

In contrast to these many clearly dualist references in the Writings, there are still hints of an underlying monist view. As we saw with Gen 2:7, these books refer to the creation of humans with a confusing mixture of monist and dualist leanings. Job refers to himself being formed from clay (Job 10:8–9), and the psalmist to humans being knit together in the mother's womb (Ps 139:13–16), without any reference to the infusion of an immaterial component (such as the breath of life). However, this passage in the Psalms confuses the picture by claiming that, in addition to knitting a very material body in the mother's womb, YHWH also "created my inmost being . . . your eyes saw my unformed body." Is this a reference to an immaterial component; something which God could see even though it did not have form? Furthermore, the location for this creative act is "in the secret place . . . in the depths of the earth" (v. 15). Might this curious reference to "the depths of the earth" trace its origin to the ancient Mesopotamian belief (the context in which these Hebrew texts were written) that the unborn child (the *kūbu*) lived in the underworld until it transitioned to the mother's womb?

The Writings also exhibit a schizophrenic double-mindedness regarding the concept of resurrection: the authors of Job, Ecclesiastes, and the Psalms seem on the one hand to deny the possibility of resurrection (Job 14:7–12; Eccl 3:18–19; Pss 6:5; 30:9; 88:11–12; 115:17) but then elsewhere seem to open up this possibility (Job 19:25; Eccl 12:7, 14; Pss 16:9–10; 49:15; 73:23–26). We will see in chapter 4 how this possibility is then very clearly and distinctly affirmed by Daniel (Dan 12:1–3, 13).

These writings, then, reveal a confusing period when thinking on this question of human ontology was in flux: when a relatively ancient monistic view was pushing back against a strengthening dualistic view, and a very new perspective on the afterlife was beginning to emerge (including the concepts of resurrection and postmortem judgment). It is noteworthy that the eight occurrences of the term for an inhabitant of Sheol (*r'pā'îm* or *rephaim*) are found only in Job, Psalms, and Proverbs, as well as in Isaiah (generally dated as exilic or postexilic).

The New Testament

The trend toward a dualist perspective continued as Judaism became Hellenized; this development can be seen clearly in many apocryphal and pseudepigraphic texts (the *Book of Jubilees*, *Sibylline Oracles*, *2 Enoch*, and *Tobit* in particular address the creation of humans, the fall and the afterlife), and in the writings of Philo. Likewise, early Christianity developed within a very Hellenistic setting and was unpacked by Greek-trained thinkers. Many Greek philosophers held a very distinctly dualist view: many of them saw the immaterial and immortal soul seeking to escape the material and mortal body in which it was trapped, and to rejoin the immaterial *nous*.

It is easy to point to New Testament passages which seem to separate material and immaterial components. The Synoptic Gospels quote Jesus echoing the Mosaic command (Deut 6:5) to love God "with all your heart, soul, and strength (Matt 22:37; Mark 12:30; Luke 10:27). The Gospels of Mark and Luke add the word "mind," while Matthew substitutes the word "strength" with "mind." Some might claim that Jesus actually viewed heart, soul, and mind as the same thing and was merely using such redundancy as a tool for emphatic effect (referred to as parallelism). However, would one say the same about my suggestion that "you need to pick that thing up with your fingers, and pass it to me with your thumbs"? Is such wording not a bit more confusing—simply because it sounds very dualist, and seems to imply two different functions or actions—than simply twice using the more monistic word "hand"? Attributing the redundancy in these gospel passages to parallelism is indeed one interpretation: another interpretation is that Jesus saw these things as being somehow different and distinct. Elsewhere, Jesus warns his listeners to "not be afraid of those who kill the body but cannot kill the soul. Rather, be afraid of the One who can destroy both soul and body in hell" (Matt 10:28).[13] Jesus scolds the disciples for falling asleep during his hour of greatest anguish, and says, "The spirit is willing, but the flesh is weak" (Matt 26:41). The gospel accounts of the crucifixion tell us that Jesus "gave up his spirit" (Matt 27:50; Mark 15:37; Luke 23:46; John 19:30), and the writer of Acts uses the same phrasing for Ananias and Sapphira (Acts 5:5, 10), and for Herod (Acts 12:23). Paul often juxtaposed references to the body against those referring to the heart, soul, and mind. He also frequently wrote in a way that seemed to separate out body from the soul (1 Cor 5:3, 5; 2 Cor 4:16; 1 Thess 5:23): he looked forward to being "away from the body and at home with the Lord" (2 Cor. 5:8); he prescribed that the Corinthian church should "hand this man over to Satan for the destruction of the flesh, so that his spirit may be saved on the day of the Lord" (1 Cor 5:5); he wrote about how we exchange our earthly body for a new heavenly one (2 Cor 5:1–4).[14] James made reference to the body without a spirit, and the animating effect that the human spirit has on the body (Jas 2:26). Peter wrote about Jesus being physically dead but "made alive in the

---

13. Green provides an alternative interpretation of this passage: Jesus was distinguishing between the end of life in this world (martyrdom) versus the end of one's entire human existence (Green "Bodies," 162).

14. Green explores the interpretation that Paul is not dualist in his thinking here, but rather is using dualist imagery of ancient Israel transitioning from a tent-based worship to a temple-based worship as a context for the transition which Christians can look forward to in moving from an earthly based body to a heavenly based body, both bodies being holistically tied to the soul (Green, "Restoring the Human Person," 20).

Spirit" and preaching to imprisoned spirits (1 Pet 3:18–19). The writer of Revelation uses Greek sentence structure which clearly distinguishes between bodies and souls when writing about the merchandise being traded in the city of Babylon (Rev 18:13).[15]

Some passages in Scripture seem to not only distinguish between material and immaterial, but even separate out the inner component into two parts: the soul versus the spirit (Isa 26:9; Luke 1:46–47; 1 Thess 5:23; Heb 4:12): the latter passage in particular portrays the soul and spirit as distinct entities. Other passages speak of God's own spirit becoming part of our being (John 20:22; 1 Cor 3:16), which introduces even more complexity in the makeup of our inner being.

Synthesis

Altogether, then, I myself do not find a consistent picture vis-à-vis monism versus substance dualism within the three lines of evidence that I explored here, though there is a strong leaning toward dualism:

i. The creation narrative of the first man opens with a dualist image of God combining material and immaterial components, but concludes with a monist image of those parts becoming a unified *nephesh*. The latter view is reinforced when the first female *nephesh* is created simply from the man's rib, which presumably was sufficient to also confer a consciousness or soul. Likewise, the creation passage in Psalm 139 employs a confusing mixture of monist and dualist imagery.

ii. Passages referring to the moment of death also first present a monist view which then seems to evolve into a dualist one. That is, Genesis tells us the person simply returns to the dust without mentioning any kind of continued existence (consistent with monism), but later refers repeatedly to a postmortem transition to Sheol (leans strongly toward dualism). Later texts present an increasingly dualist view, and the latter becomes unavoidable within the exilic texts and then the New Testament.

---

15. Green, "Bodies," 150; Green, *Body, Soul*, 47. This distinction is evident in translations such as the New Revised Standard Version and the Living Bible, but lost in others such as the New International Version. This might represent an example of the theology of the translators influencing their interpretation of the texts, and then those interpreted texts being used to corroborate the theology of their readers (a form of circular argument).

iii. Passages which describe resurrection and/or the after-death existence also increasingly point to dualism in that it is clear that the immaterial personhood continues long after the material body has decayed away. The earliest texts paint an image of a gloomy nonexistence in Sheol, cut off from loved ones and from God himself (while at the same time confusingly referring to being gathered to one's people and "sleeping" with one's fathers). However the disembodied *rephaim* are increasingly seen to be interactive with the living (King Saul's summoning "Samuel"; Israel's development of a death cult) and able to return from the Land of No Return. Finally, the belief that the faithful can look forward to leaving behind a persecuted, martyred body and receiving some new kind of body as part of their reward which they enjoy in a celestial setting leaps out of the pages of Daniel and the teachings of Jesus and Paul.

As such, it is imprecise to say that the biblical view of human ontology is monolithic, let alone that it is decidedly monistic in nature: it does indeed portray the person functioning as a psychosomatic unity as long as the body is alive, but it clearly teaches that something—presumably immaterial, and which conveys the very personhood of that individual—leaves behind the physical body at the moment of death and continues to exist in some other manner or dimension. This is entirely consistent with the predominant view among most modern lay-believers that humans comprise a duality, having an immaterial soul separate from (but residing in) the material body.

This analysis shows that there has been a tension between monist and dualist thinking throughout the writing of Hebrew and Christian Scripture. Irrespective of whether one sees the biblical writers as monists or dualists, the early Christian church quickly became *decidedly* dualist under the leadership of her Greek-trained shepherds: they saw humans comprising a duality of substances, the body and the soul/spirit, and developed a very elaborate theology around this. Gnosticism, which was opposed by Christians as being heretical, was very strongly dualistic in its theology (not just soul versus body, but also light versus dark, good versus evil, truth versus falsehood) and actively pursued strict asceticism (punishment of the body) and practices intended to liberate the immortal soul from the prison of the body.[16] Strong dualism was rearticulated and strengthened by numerous Christian scholars a millennium later yet. As already noted above (page 62), modern scholars within the academy may have returned to the monist position, but the vast majority of Christians in general today seem to be quite dualistic in their theology, and most non-Christians seem to believe that a

---

16. Anderson, "On Being Human," 183–84; Barbour, "Neuroscience," 252.

dualist worldview is central and essential to Christian theology.[17] "So much of Christian anthropology is anchored to a dualist narrative that any other rendering of the human person might seem to shake one of the main pillars of Christian faith itself."[18]

Weaknesses of a Dualist Framework

Scholars have provided excellent summaries of the shortcomings of earlier forms of substance dualism, as well as recent revisions which attempt to address physicalist criticisms of them.[19] The first criticism is that dualism often only injects a component which asserts but does not explain. It is essentially a "spirit-of-the-gaps" argument: "I know it does, just because it does. I don't know how else to explain it, so that's where I stand." For many scholars, this form of reasoning is unacceptable. It suffers from the same vulnerability as the God-of-the-gaps argument: as neurobiology increasingly explains all the properties and functions that we have traditionally ascribed to the soul (memory; emotion; volition; reason; decision-making; etc.), the soul increasingly becomes just an epiphenomenon.[20]

Another major weakness of any model which includes a separate immaterial component is the question of precisely how does the immaterial *cause* a change in the material realm. This question has long vexed Christian thinkers: this is easily traced back to Descartes and the seventeenth century, but one can find this pebble in the shoes of Augustine in the fifth century and even Athanasius of Alexandria in the fourth. I have mentioned above how Descartes was inspired by the technology of the church organ to posit that something immaterial within the brain exerted hydraulic pressures on the muscles through the spinal cord and nerves which emanated from it, and further posited the pineal gland as the "organist" playing the keys on the "organ."[21] Modern science confirms that Descartes was correct in pointing to the brain and nerves as controlling muscle movement, but corrects him by stating they do so using electrical and chemical signaling rather than

---

17. Murphy, "Human Nature," 22, 24; Peters, "Resurrection," 306–7.

18. Green, *Body, Soul*, 13–14

19. Cortez, *Theological Anthropology*, 68–97; Shults, *Reforming Anthropological Theology*, 172–84; Scott and Phinney, "Relating Body and Soul," 90–107; Brown, "Neurobiological Embodiment," 58–76; Green, *Body, Soul*, 13.

20. Green, *Body, Soul*, 27.

21. Cortez, *Theological Anthropology*, 75; Shults, *Reforming Anthropological Theology*, 174; Ward, *Big Questions*, 143; Murphy, introduction to *Neuroscience and the Person*, xiii; Dennett, *Consciousness Explained*, 34.

hydraulic pressures. But this scientific update still does not answer the central question. It still provides no description of the interface between the material and immaterial: what exactly is the "organist" and how does it exert a material effect (whether that be hydraulic pressure or an electrical signal)? Descartes's defenders might have pointed to the steam engine (which was invented around the time of Descartes's death) as a useful concept or even precedent to address this conundrum: even a formless water vapor which is so delicate that it can be easily blown away from one's cup of tea by a child's breath is nonetheless capable of moving a one hundred ton train or ship given the right machinery to transduce its effect (that is, the steam engine). And even today, some readers might recognize that this ultimately fails as an *explanation* but still feel it serves as a useful *analogy* toward finding the missing "organist": however, they should take note that everything in that analogy is *material*, including the steam (heated water molecules) which is meant to stand in for the immaterial spirit. It still does not provide any clarity to how the *immaterial* might move the material. Magnets can be used to exert forces on objects which contain iron or nickel, but have no effect on things made of aluminum or carbon. Likewise, a spirit is immaterial and therefore does not interact with matter, so how would it physically move the first sodium and calcium ions in order to create an electrical signal which activates the first nerve ending? One can *just believe* that they do, but that again amounts to a spirit-of-the-gaps explanation. And why would there be such an exclusive pairing between one soul and one body: that is, why would the immaterial component of one person be only capable of manipulating the calcium and sodium ions of the body in which it "resides," but be completely unable to manipulate those in another person's body, even during an intimate embrace?

Some also rebut that any intervention from outside of the closed system of the material realm represents a violation of thermodynamics: more precisely, the conservation of energy.[22] That is, any material change would require expense of energy. Sir Isaac Newton shuddered at the amount of energy being expended by the angels who were carefully shepherding the moon and planets through their orbits every day for millennia, like Sisyphus forever rolling a stone to the summit of a hill, only to have it roll down the other side, requiring him to start all over again. Eventually, however, Newton found a way to substitute the immaterial angels with various material laws of motion and gravitational attraction *which were internal to the closed system* and which could account for those orbits. Likewise, if some spiritual force were always needed to mobilize the first sodium and calcium

---

22. Peters, "Resurrection," 313; Dennett, *Consciousness Explained*, 35.

ions behind every thought of every human, an immense amount of energy must be constantly injected into the closed system of our universe. This would break the laws of thermodynamics, and some question whether God needs or chooses to break laws of science any more than he might break laws of logic (for example, by creating a circle that is square, or a rock that is too big for even him to lift).

More recently, Scott and Phinney introduced several other insights gleaned from brain sciences which were totally unavailable to Christian thinkers prior to a few decades ago, and which further challenge a dualist explanation of human ontology.[23] In particular, developmental biology highlights theological difficulties which arise from certain special cases of embryogenesis. Depending on one's view regarding precisely when the fertilized egg receives a soul, one may struggle to explain how a second additional soul is provided when an embryo splits and produces identical twins, or what happens to the "extra" soul when two genetically different embryos fuse to produce a single fetus. Alternatively, for both special cases, does one solve the problem by asserting that the human soul is given at some later stage in embryonic development, which only introduces the same conundrum that has long dogged the abortion debate: at what point do we recognize the fetus as being human?

Dualism can also create a philosophical or ethical paradox, especially for those who associate the soul with the brain. When a portion of human brain tissue is transplanted into another recipient's brain, as has been done in the past as a potential therapy for degenerative brain diseases like Huntington's disease,[24] does the donor's soul accompany that piece of brain into the recipient? If one's answer to that question makes reference to the relatively small size of the implant, they might be asked to declare where the threshold would be at which the implant is "big enough" for this to be a legitimate question. If that piece of brain tissue is instead clonally expanded and distributed to different laboratories around the world, which again has often occurred, are those scientists essentially creating multiple souls? If science ever reaches the point at which it can use cell cloning, stem cell biology, and tissue engineering technology to create an entire human body out of that piece of donated brain tissue, what soul would that clone have?

Finally, Nancey Murphy raises an entirely different theological concern. "I take dualism to be theologically undesirable because of its penchant for distorting Christian priorities. That is, it has given Christians something

---

23. Scott and Phinney, "Relating Body and Soul," 90–107.

24. Freeman et al., "Neural Transplantation," 405–11; Peschanski et al., "Rationale," 273–85.

to be concerned about—the soul and its final destiny—in place of Jesus' concern with the kingdom of God."[25] This echoes a warning that others such as N. T. Wright have often voiced: that modern Christianity is more concerned about how to leave earth and get to heaven than how to bring heaven to earth.

These and other scientific and philosophical arguments have begun to move the scholarly community away from a fully substance dualist worldview (although, as already mentioned more than once, the church as a whole has not followed them: most of her members are still quite dualist in their own personal theologies). Those scholars may fully agree that humans are not simply biological machines, but they nonetheless struggle with (or have philosophical prohibitions against) simply adding to that a mysterious, immaterial, immeasurable, undefinable, and unexplainable component: that which I have previously referred to as a "spirit-of-the-gaps." They would seek to find an alternative view or explanation of human ontology which agrees with modern scientific discoveries about the brain and cognition without violating Scripture.

One way to resolve the tension is to soften the dualism. The earlier forms of substance dualism presented the material and immaterial as entirely and completely distinct and separate, whereas contemporary dualists see them as conceivably separable but still a single functioning whole. Cortez presents three views on this:[26]

1. Holistic dualism: two distinct parts which are "fully integrated and interdependent such that the organism as a whole functions properly only when both are working in intimate union" (an analogy might be the computer hardware and the software which it runs).

2. Emergent dualism: the mind is an emergent *substance* of the brain (in contrast to the physicalist explanation that the mind is an emergent *property* of the brain, which we will explore in later sections of this chapter).

3. Thomistic dualism: building on Aristotelian considerations, the soul is a Form of the body. That is, all material objects comprise a material composite (the matter from which it is composed) and a substantial form (determines the essential nature of the object).

Cortez acknowledges that to the same extent that one holds onto any form of dualism, one keeps oneself open to the criticism of also holding

---

25. Murphy, "Nonreductive Physicalism: Philosophical Challenges," 97.
26. Cortez, *Theological Anthropology*, 21, 72–75.

onto "mystery" in one's worldview, something which pure scientists find unsettling. Cortez pushes back by saying that physicalists, despite their protestations regarding the mystery of consciousness, are nonetheless comfortable with mystery vis-à-vis free will and creativity (we will elaborate upon this below on page 118) or in other scientific areas (for example: multiverse theory; quantum mechanics; string theory; the definition and origin of life). As such, given the choice between a monist/physicalist worldview versus a dualist one—both leaving many questions unanswered and leaving much unexplained as simply mystery—the dualist position is simply preferable to Cortez.

Scott and Phinney have attempted to push the ball a little further by introducing a model—"developing hominization"—which they feel better addresses the shortcomings of other dualist proposals.[27] In particular, they suggest that those faculties which we attribute to the human soul are products of an "essence" (a substance or property) which "interacts so intimately with the entire person, that it is only the entire person that exhibits functional unity." This essence is not present in its complete form at fertilization, and grows in its manifestation as the fetus develops. Furthermore, God can maintain or recreate this essence even after death and decay of the body. It is not clear (to me) how such a model differs from Tertullian's third-century strictly dualist model in which the immaterial soul is created at conception with the fertilized egg and develops along with the person's body, itself going through its own developmental stages including a spiritual form of childhood and a "puberty of the soul" (page 54). In Tertullian's mind, the soul continues to develop in response to the religious experiences and spiritual practices of the person, ideally attaining full maturity, but ceases to continue developing further when absented from the body: the arrested soul and the recreated body are later reunited at the Eschaton. It seems that the primary difference between Scott's and Phinney's idea and Tertullian's dualist view is that the soul is softened to an "essence" rather than a substance.

In the end, attempts at responding to the criticisms against substance dualism have merely amounted to softening the term "substance" or producing a fairly dualistic form of monism. On the other hand, other scholars have found it possible to explain many of the capabilities and features of consciousness, mind, personality, and even the soul using fully naturalistic and physicalist mechanisms operating within the human brain. In the next few sections, we will explore several such physicalist explanations. Some readers may feel threatened by this. They may fear that such a move can overemphasize reliance upon physicalist explanations and a tendency

---

27. Scott and Phinney, "Relating Body and Soul," 99–103.

toward the dismissive attitude of people such as Francis Crick, co-discoverer of the molecular structure of DNA and staunchly vocal atheist, who declared "you are nothing but a packet of neurons."[28] Clayton has asked: "If neuroscience is successful in this way [describing personhood in purely physicalist terms], will we have *proven* that all things that exist are physical, or that the conscious self is an illusion?"[29] He then replied to this: "By no means! To move from successes in neuroscience to the doctrine that only physical things exist—whether one then advocates the falseness of belief in God or interprets the self as 'merely metaphorical or constructed'—is a category mistake." The reader is encouraged to continuously resist Crick's extreme end of the spectrum of reductionism: Christian scholars in this area of research insist that it is indeed possible to find a *non-reductive* physicalism which does not take this radical step nor force one to give up a fully Christian theology. However, it *will* require some reexamination of certain theological assumptions and conclusions.

Mind and Personality as Emergent Properties of the Brain

In the cognitive sciences, "emergence" refers to the appearance of something new and completely different from the material or immaterial things which originate it. Cyberspace is an example of an emergent phenomenon: it is not an actual, material *place* anywhere in this universe, but a *virtual reality* which arises through a complex amalgamation of computers, servers, vast lengths of electrical and optical cables and software. Another beautiful example of the concept of emergence in material things is the phenomenon of a murmuration of birds, in which many tens or hundreds of thousands of starlings or warblers or other birds swoop in a billowing cloud that takes on characteristics of an organism with a form and mind of its own (a similar phenomenon is found in large schools of fish, or large herds of certain mammals). Readers who are not quite certain of what I am describing here are strongly encouraged to search for the phrase "murmuration of birds" using YouTube: prepare to be amazed. It is so hard to not see that flock as one large organism which is itself completely unlike a bird. And yet it is *not* a large organism: it is simply a large flock of individual birds, each one acting individually but obeying a common set of rules which have been preprogrammed into them. In fact, researchers have studied this behavior and reduced it to three simple rules: the birds are constantly assessing the distance

28. Crick. *Astonishing Hypothesis*, 7.
29. Clayton, "Neuroscience, the Person, and God," 190.

between themselves and their immediate neighbors (believed to be roughly six or seven) as well as the average distance and heading of the entire flock, and then steering: (1) toward the average heading of the flock; (2) toward the average position of the flock; (3) away to avoid collisions with crowding neighbors.[30] These three simple rules alone can be programmed into a child's computer game which then completely simulates the phenomenon of an avian murmuration. Even immaterial things such as ideas, behaviors, or music can be brought together to form something entirely new and different; simple mathematical formulae, when plotted against a pair of coordinates, can produce amazing patterns that are simply not apparent within the formulae themselves. In fact, a popular game from my own childhood era—Spirograph—used colored pens guided by interlocking gears to create stunning geometric drawings which could not be predicted before pen was put to paper.

Many now see this concept of "emergence" as being useful in explaining various aspects of our mind and soul. It is clear that the brain is centrally involved in many (all?) aspects of our inner being. Using sophisticated imaging techniques such as magnetic resonance imaging (MRI) and positron emission tomography (PET), it is possible to document the activation of discrete brain regions engaged in a wide variety of tasks: processing sensory information; facial recognition; planning and executing a movement; abstract or deep philosophical thinking; meditation; dreaming.[31] The use of language derives from the semantic capacities located in a complex of regions in the brain including Wernicke's Area and Broca's Area. The visual cortex is responsible for collating visual sensory input together with a wide range of other inputs to produce a mental representation of the world outside of our brain. The movements of specific parts of our body are choreographed by discrete brain regions laid out in anatomically sequenced sections of our motor cortex (referred to as a homunculus).

Those brain regions and pathways all employ "integrate-and-fire" neurons. That is, each neuron has a central body from which extend varying numbers of dendrites (which *receive* input signals from other "upstream" nerves) and axons (which *send* the output signal of that nerve to other "downstream" nerves); figure 2 on page 97 will give the reader a mental image of this. A chemical signal—a neurotransmitter—is exchanged at the contact point—or "synapse"—between two nerves, which binds to a specific receptor on the postsynaptic nerve: that binding may or may not cause electrical excitation of the region of membrane immediately around the receptor (depending on

---

30. Ballerini et al., "Interaction Ruling Animal Collective," 1232–37.
31. Clayton, "Neuroscience, the Person," 182–84.

a variety of factors). The electrical signals created in all of the dendrites and even on the nerve body itself add together: if the integrated signal is sufficiently high, the axon will "fire" and release its own neurotransmitter signal to the next neuron(s) in the signaling pathway.

Putting all these concepts together, it is thought that as one brings together increasing numbers of neurons and/or creates enough connections between them, intelligence and a personality will begin to emerge out of the signal processing occurring within and between them. This is not at all preposterous: once again, we have already seen this reduced to simple rules and then reproduced in computer programs. Our first computer programs designed to "talk" with humans were quite clunky, and followed very narrow preprogrammed scripts. Today, however, one can often find oneself making a phone call to a large business or government office and actually "talking with" a computer that asks questions, interprets our spoken responses despite a myriad of potential linguistic accents and speech defects, and collects information from us to intelligently direct us to the right department or representative. This too has become so commonplace that it has been put into the hands of children: today's Siri, Cortana, Alexa, and Google are truly astounding. It has all come about by simply adding increasing numbers of lines of computer coding and modules which handle specific computing tasks. Many believe that as computers become more complex and incorporate ever increasing numbers of processing units and algorithms (analogous to neurons and synaptic connections), it will be possible to re-create *in silico* something akin to the human mind and thereby produce artificial intelligence.[32] There are even some who envision a future in which the unique synaptic connections within each of our brains can be analyzed and reproduced within lines of computer coding, and in so doing recreate everything that defines the mind of a person. In other words, they believe it will someday be possible to be "uploaded" into a computer and thereby live free of the limitations of our bodies (a modern redux of Neoplatonism?). Needless to say, this raises all kinds of questions, problems, and conundrums.

Our computers are not the only evidence we have for intelligence arising out of complexity. We also see this in nature itself: animals sharing to varying degrees certain cognitive abilities which we attribute to the soul. We now have abundant (and still accumulating) evidence of animals exhibiting emotion, volition, creativity, a sense of self, and theory of mind. Few dog owners will deny that their pets feel emotion (the same might be said of those who own horses, cats, and other kinds of pets), and they may also adamantly claim that their pets know when they have broken a house rule. Likewise,

---

32. Bjork, "Artificial Intelligence," 95–102.

primatologists have shown that chimpanzees and bonobos are aware when a rule has been broken, or when something has been "unfair": that is, they seem to understand the difference between "right" and "wrong," or "good" and "evil," bordering on a sense of ethics and morality. Chimpanzees, bonobos, capuchin monkeys, macaques, bottlenose dolphins, sea otters, and many species of birds are able to make and use tools.[33] Many birds can be quite creative in the way that they decorate their nests, and elephants can paint pictures: even though the latter may be a trained behavior which would never occur in the wild, the exceptional quality of their paintings does indicate some kind of sense of abstract thought, appreciation of form and an artistic ability to reproduce the images they see. Animals in general employ a diverse range of forms of communication (their own unique "languages"), and some—such as chimpanzees, bonobos, and orangutans—seem to do so in a way that is uncannily similar to that of humans (especially in the form of sign language). Koko the gorilla was said to have learned to use one thousand hand signs based around the American Sign Language system, and understood roughly two thousand English words. Some critics minimize these facts as unremarkable, accusing her of not using syntax (she might indicate "Koko banana hungry" rather than "I want a banana"), and that her language level was on par with "only" a young human child. It's a shame those critics miss the bigger picture: they're still comparing Koko to a human!

It is becoming increasingly apparent that our minds and personalities are tied directly to the actions of individual neurons (akin to the responses of a murmuration of birds being tied to those of the individual birds). Neurobiologists can point to specific brain anatomical locations that seem to be directly responsible for a wide range of cognitive abilities like emotion, language, sensory perception, volition, and memory. These abilities certainly pertain to our mind and personality; we will see below that they also pertain to our soul. This attribution is based upon our ability to experimentally elicit or modulate those faculties by simply stimulating discrete areas of our brain electrically or pharmacologically, or by observing the changes in people with precisely defined neurological defects and brain injuries.[34] An often-quoted example of the latter is Phineas Gage, a railroad worker in 1848 who suffered a traumatic brain injury as a result of an explosion which drove a large spike through the front part of his brain.[35] Despite losing large sections of his frontal cortex, he made a relatively speedy and nearly complete

---

33. Wilcox, "Proposed Model," 22–43; Roffman et al., "Varied Natural Tools," 78–91.

34. Jeeves, "Neuroscience," 173.

35. Parkes, "Vulnerability," 47–48; Ward, *Big Questions*, 149; Shults, *Reforming Theological Anthropology*, 179; Jeeves, *From Cells to Souls*, 47; Jeeves, "Neuroscience," 174; Brown, "Cognitive Contributions," 121; Green, *Body, Soul*, 82.

biological recovery and was able to rejoin society, but his personality was forever altered: all who knew him found he had become a completely different person: "Whereas Gage's intelligence itself did not seem to change, the considerable wisdom he had possessed as a manager of workers and responsible member of a community and family were lost. Important qualities of Gage's personal relatedness were lost or altered."[36] Brown recounts a similar story in the modern context of an individual whose personality was radically altered following surgical excision of a frontal lobe meningioma, even though all other cognitive aspects (intelligence; sophistication; social judgment; memory) appeared to be entirely unchanged.[37] In another story, bizarre and disturbing behavior in an individual disappeared after a brain tumor was surgically removed, but then reappeared at the same time that the tumor reappeared, then subsided again when the regrowth was removed.[38] Many neurodegenerative diseases such as Alzheimer's can bring on dramatic and disheartening changes in personality—to the point that family members can sometimes lament "this is not the loved one we once knew"—even though the only thing that seems to have happened is that neural connectivity has been decreased (rather than some portion[s] of their spirit/soul having departed).

Other brain diseases and injuries can in fact produce multiple personalities. Many readers will be familiar with this happening in schizophrenia, but may not be aware that injuries which eliminate the communication between the left and right hemispheres (referred to clinically as a "split brain") can lead to a situation in which the person exhibits signs of having two distinct minds. For example, Barbour writes about how "split brain subjects will carry out an action with one brain hemisphere that uses information from a visual input to the other hemisphere of which they are not aware; they will then try to explain their action by some other reason, unrelated to the visual input."[39] (Through a quirk of our brain circuitry, the left half of our brain receives visual input from only the right side of our peripheral vision, and it controls the right side of the body, while the opposite is true of the right half of our brain.) In other experiments, split-brain patients were asked to solve a problem which required logic and spatial reasoning: they could do so using one hand alone, but not the other hand alone, and when both hands were allowed to be involved the two hands competed with each other as if controlled by two different people. Most readers themselves have almost certainly often had forms of "split-brain" experiences: how else

---

36. Brown, "Cognitive Contributions" 121.

37. Brown, "Cognitive Contributions," 121–22.

38. Burns and Swerdlow, "Right Orbitofrontal," 437–40; also see Scott and Phinney, "Relating Body and Soul," 99; Green, *Body, Soul*, 73.

39. Barbour, *Neuroscience*, 261; also see Gazzaniga, "Brain Modularity."

does one explain being surprised by something completely unexpected in a dream (since it was obviously her own brain, or at least part of it, which conceived in great detail that monster which suddenly sprang up from around the corner)? Could this also explain, at least in part, one talking to oneself when in deep mental thought or problem-solving and in so doing suddenly discovered a new perspective on the situation at hand? Could this explain how one can simultaneously hate oneself while still loving oneself?

Not only does altered brain structure lead to altered behavior, but there are also examples of altered behavior altering brain structure. For example, MRI studies of London taxi drivers show significant increases in the volume of their hippocampus (which stores spatial representations of the environment), and the increase in volume correlates with the amount of time spent as a taxi driver.[40] Similarly, there is evidence that the loss of cognitive function seen in neurodegenerative disease, and even in aging, can be held back by cognitive stimulation (such as doing word puzzles), analogous to the increased muscle mass produced by physical exercise.

Altogether, these observations can be difficult to explain if the mind or personality is not tied directly to the neurons of our brain (for example, if it is believed instead to be an emanation from the *nous* or a separate immaterial entity). It is no stretch to acknowledge that many (all?) aspects of our mind and personality derive directly from the signal processing and neural integration of our brain. Many of the functions of the mind are subconscious and involve input/output pathways that can be fairly precisely defined and even modulated. But what is "consciousness" itself, or the "soul" or "spirit"? These are the questions to which we will now turn.

Physicalist Explanations for Consciousness?

Some physicalists claim there is nothing to explain when speaking of consciousness: for them, consciousness is an illusion that humans have contrived to give meaning to something which does not exist beyond being simply a state of mind. It is not only not material (the double negative here is important), it is not even a "thing" to speak of. The main reason they think this way is that they feel science is the only path to truth, and consciousness appears to be something that is difficult (they might say impossible) to measure, handle, or study by scientific methods. It is worth pointing out to those thinkers that claiming "science is the only path to truth" is itself not a scientific statement but a purely philosophical one: a worldview. In this, they are indulging in blatant question-begging, although they may not see it. Such dyed-in-the-wool physicalists choose to only use physicalist methods and

---

40. Wollett and Maguire, "Acquiring 'the Knowledge,'" 2109–14.

technologies and accept only hard physical data, and are thereby seeing their world through metaphorically rose-colored glasses: physicalist approaches will only be able to probe physicalist properties of an object, but that does not mean that the object has *only* those properties.[41]

That having been said, others have indeed found some explanatory power for consciousness from physicalist approaches. These see consciousness as an emergent property of neural signal processing. No one has succeeded yet in providing a complete theory for consciousness: all the models and theories have problems. Attention Schema Theory and the Integrated Information Theory propose that the brain has been evolving increasingly complex and layered mechanisms to focus its signal processing resources on a select few signals at the expense of many other ones less important for survival, and eventually led to consciousness in the vertebrate lineage.[42] Others propose the Global Workspace Theory, in which consciousness involves a diffuse broadcasting of information across the entire brain.[43] Dennett holds out a Multiple Drafts model, in which the conscious mind is continuously integrating external (sensory) and internal (processing) inputs and developing plausible representations of reality.[44] Some of these scholars would then posit that our stream of consciousness constitutes the moment-by-moment acceptance or endorsement of the series of computational integrations from these processes, much like the individual snapshots in a long series of picture frames which are stitched together in one long loop of movie film. Others among them, Dennett in particular, reject the use of words like "acceptance" or "endorsement," as well as any notion of a "stream of consciousness," because they disallow the idea that there is a "Cartesian theater" in which some kind of immaterial substance is doing the accepting, endorsing, and experiencing.[45]

All of the proposals mentioned above focus on the cognitive strategies employed by the brain to produce consciousness, but do not really point to cellular or molecular mechanisms (how consciousness is "done," versus exactly what is "doing" the consciousness). The famed astrophysicist Roger Penrose and anesthesiologist Stuart Hameroff have recently proposed that a structural molecule in our neurons—tubulin—might account for consciousness itself.[46] I will here provide my summary of their hypothesis, but

---

41. Clayton, "Neuroscience," 198.

42. Webb and Graziano, "Attention Schema Theory," 500; Oizumi et al., "From the Phenomenology," e1003588.

43. Baars, "Global Workspace Theory," 45–53; Baars and Franklin, "Architectural Model," 955–61; also see Dennet, *Consciousness Explained*, 257.

44. Dennett, *Consciousness Explained*, 111–38.

45. Dennett, *Consciousness Explained*, 253–55, 257.

46. Hameroff, "How Quantum Brain Biology," 1–17, 93; Schwartz, "Quantum

will warn the readers that this is going to be a highly scientific section. I have worked hard to reduce the concepts to their most basic elements and used analogies to help the reader form a picture of what is going on. Just the same, this will test the limits of some readers. To summarize this entire section in just one short pithy sentence: these investigators propose that a molecule in our nerve cells called tubulin might be at the heart of a mechanism that produces consciousness using chemistry, biochemistry, standard/classical physics, and quantum physics. That may be enough for readers who prefer to avoid scientific details, and they may now choose to jump directly to the next section about spirituality. The more confident and adventurous readers may instead choose to wade more deeply into the details of this tubulin-model of consciousness. Hang on tight!

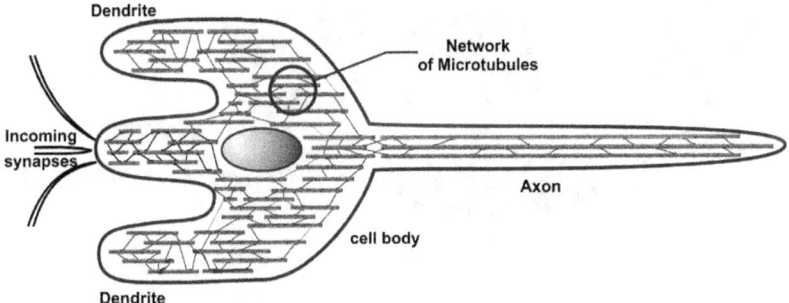

Figure 2

A cartoon diagram of a neuron, featuring three dendrites pointing out to the left of the central body, and one axon pointing out to the right. The dendrites receive information from "incoming synapses" of other nerve endings, and send information out via the axon to other nerve cells (via "synapses"). Inside the neuron is a network of microtubules which form the skeleton of the cell (see fig. 3 for more detail of these microtubules). Image modified from Hameroff and Penrose, "Consciousness in the Universe"; with permission of the author.

All cells, including neurons, have an internal skeleton made up of tubulin fibers referred to as microtubules (see figure 2); one can envision this skeleton looking like the poles and wires which hold up a circus tent (in this analogy, the canvas tent itself represents the outer membrane of the cell). The individual tubulin proteins have different patches on their surface with chemical or physical properties which help them to link up end-to-end and

---

Physics," 1309–27; Korf et al., "Quantum and Multidimensional," 345–55.

side-by-side to form into long beams and crossbeams called microtubules (the individual proteins are referred to as monomers, while collections of tubulin monomers bound up together are referred to as polymers).

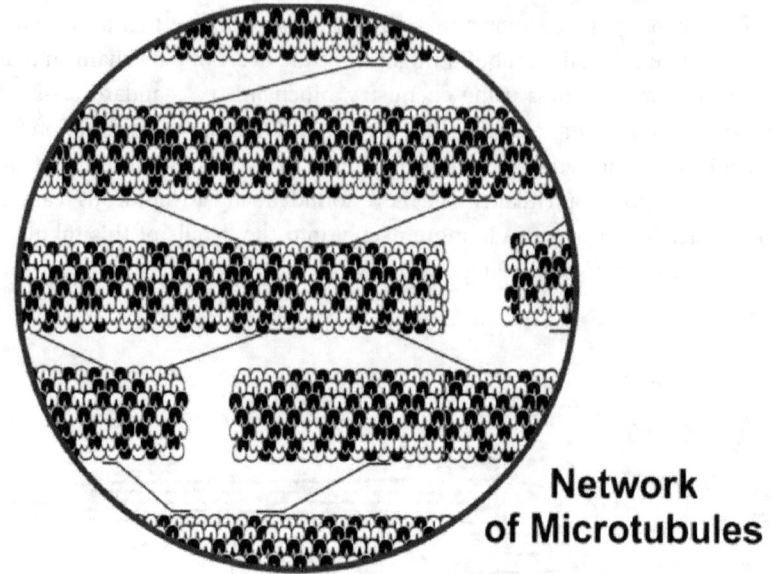

Figure 3

Magnified image of microtubule network enclosed within circle in figure 2. Individual tubulin monomers (here colored either white or black) assemble into large "beams" of tubulin polymers, which in turn are linked by slender cross-beams made of other proteins. The white/black coloring of the tubulin monomers can be taken to represent the different subtypes of tubulin monomers (encoded by different tubulin genes, or following biochemical modifications such as phosphorylation). That coloring can also be taken to represent different quantum sub-states which influence the functionality of those tubulin monomers and polymers. Both contexts are explained in detail within the text. Image used by permission of the authors.

Figure 3 provides a magnified portion of figure 2 to emphasize the individual tubulin monomers (colored either white or black) which make up the tubulin polymer (also called a microtubule). Three of the more general of these chemical/physical properties can include:

- electrostatic interactions (a positively charged region on one monomer will bind strongly with a strategically placed negatively charged region on an adjacent monomer);

- hydrophilic or hydrophobic interactions (two regions on adjacent monomers will "want" to create a small microenvironment trapped or enclosed between them which is either water-loving or water-hating ["greasy"]);
- hydrogen-bonding (amino acids in two adjacent monomers will "want" to share a hydrogen molecule, thereby loosely binding the two together).

The reader should see these three chemical/physical properties as three different types of molecular glue which can hold molecules together, and which can be used in various combinations. Next we can consider the molecules that are being glued together. Tubulin monomers come in all kinds of varieties, somewhat analogous to the different types of Lego blocks (colorful interlocking plastic "bricks"), each one having a slightly different amino acid sequence which in turn can lead to slightly different physical or functional properties (for example, this can determine which of the three types of glues are effective in binding the different monomer types). The amino acid sequences are determined by unique gene sequences (a genetic process referred to as "expression"), and the expression of the different genes can be regulated differently with respect to place (neuron type; subregions of neurons) and time (before versus after a nerve firing). All those different tubulin monomers can be combined in various ways to make different types of tubulin polymers (analogous to the different types of Lego buildings, machines, or animals which children will make out of those Lego blocks), each of which will function differently, and interact with other tubulin polymers in different ways. Those functions and interactions can be further modified by attaching various chemical groups to the monomers and polymers (for example, the addition of a phosphate molecule), a process referred to as "post-translational processing." These and other biochemical events can modify the electrical interactions between the tubulins, as can physical distortions introduced by slight flexing of the molecules, which can bring together or separate different functional parts of the tubulins.

Altogether, then, there can be a nearly infinite number of combinations of physical interactions and arrangements of the tubulin polymers within each part of a given neuron; these arrangements can vary on timescales lasting microseconds to years. More to the point here, it is suggested that changes in the architecture, microstructure, and configurations of these tubulin proteins might affect neuronal function.[47]

47. In a sense like information being coded within the structure of genetic molecules, although this information would be read in a completely different manner than the mechanism by which our genes are decoded.

The brain works by passing on messages from one neuron to one or more other neurons. Sometimes the functional connection between one neuron and the next—which neuroscientists call a synapse—can be a little bit stronger, or a little bit weaker, than other connections. In fact, the strength of any given connection can increase or decrease from moment to moment, often because of signals that the one neuron has received from yet other neurons.

An analogy may help the reader to cement this model of neuronal processing in place. Hopefully the reader is familiar with a game in which people sit side-by-side in a line, each straining to hear a message whispered by the person sitting to their one side, and then turning to softly whisper what they *think* they've heard into the ear of the person sitting on their other side. The fun part is seeing how well or poorly that message makes it from one end of the chain to the other. Now, instead of the people only sitting side-by-side in a line, imagine that they are in a large overcrowded room, and each person is whispering that message to the four or five people nearest them in all directions. Meanwhile there is loud music playing in the background and some players are whispering particularly loudly, and others particularly softly, and all of them are adjusting the message they repeat at one moment in time based on what they think they have heard and overheard from the others in their immediate vicinity. You can imagine how the message can easily morph in a thousand different ways as it makes its way from one end of the room to the other under those circumstances. That is how your brain works.

Combine that analogy with the one above describing the skeleton of the neuron as being like the poles and wires holding up a circus tent, and then also introduce the concept of the interactions between the tubulin monomers and polymers constantly shifting in response to the changes in chemical and physical properties listed above. Each of the poles and wires of the circus tent (representing the tubulin crossbeams) are constantly shifting position (representing synaptic contact) and adjusting tension (representing synaptic strength). In the process, the contact that one part of the circus tent (representing the neuron) makes with an adjacent one changes a little bit from moment to moment. In the neurons, these small structural changes can subtly alter whether the neuron fires, as well as the strength of the communication bridge that the neuron makes with the hundreds or thousands of neurons with which it communicates. It has been suggested that information might be encoded within the architecture and microstructure of these microtubules. That is, permanent structural changes in the synaptic endings between neurons may be the physical substrate of memory, while dynamic changes in the tubulin skeleton might contribute to consciousness.

To complicate this proposal even further, researchers investigating the role of tubulin in consciousness also invoke quantum mechanics to explain very subtle sub-molecular changes within the tubulin structures, such as arrangements which can seem to exist as both "on" and "off" at the same time (a phenomenon that physicists call "superposition"). Readers may have encountered this phenomenon in the form of the famous "Schrödinger's cat" thought experiment, in which the cat can be considered to be both alive and dead at the same time until the box is opened and its actual state is observed. At the risk of greatly oversimplifying one of the leading proposals for consciousness, tubulin monomers and polymers can take on a wide variety of quantum superposition states in response to input from "upstream" neurons, together with quantum information passed several hundred milliseconds *backward in time* from the same neuron in question. These superposition states are periodically integrated ("added up"), causing the microtubule to take on a certain configuration and conductivity state which in turn can subtly influence the signaling machinery at the synaptic ending where that particular microtubule terminates. The integration step involves processes referred to in quantum physics as "reduction" of the quantum superpositions once the tubulin reaches a certain objective threshold, and is orchestrated by synaptic inputs and other factors. Altogether, then, this entire proposal is frequently referred to as the Penrose-Hameroff Orchestrated Objective Reduction model (Orch OR). Consciousness is the steady stream of such orchestrated reduction events (could this in part underlie Dennett's Multiple Drafts model [even though he rejects the concept of a stream of consciousness[48]]?). Furthermore, the sense that one might feel of time slowing down (as in a dire life-threatening situation, or when a highly trained athlete is "in the zone" [others might feel this when immersed in a video game]) is the collective focusing of larger fields of neural signaling pathways with their inherent microtubules and superposition states on a single cognitive effort.

It has been found that microtubules resonate at frequencies ranging from ten thousand to ten million beats per second (10 kHz–10 MHz). This leads to the suggestion that "with roughly 10^9 tubulins per neuron switching at, e.g., 10 MHz (10^7), the potential capacity for microtubule-based information processing is 10^{16} operations/second per neuron."[49] To translate this

---

48. Dennett, *Consciousness Explained*, 253.

49. Hameroff, "How Quantum Brain Biology," 5; also see Flanagan, *Consciousness Reconsidered*, 35–37, for a similar mathematical consideration of the mind-boggling complexity of neural connectivity, resulting in there being $10^{99,999,999,999,997}$ functionally unique states or configurations of the human brain, compared to only 10^{87} particles in the known universe.

very scientifically worded statement into somewhat more familiar language (hopefully): a billion tubulin molecules all shifting their configurations ten billion times each second produces ten quadrillion (10^{16}) configurations of the tubulin skeleton each second, and each configuration can subtly alter the electrical activity *of that individual neuron*. That processing speed of a single neuron (10^{16} operations/second) far exceeds that of today's fastest parallel computers. Then considering that there are a hundred billion neurons in our brain, and one hundred trillion connections between them all, one can appreciate that our brain could be an amazingly complex information processing machine that in many ways far outperforms our most sophisticated computers and artificial intelligence software. It is also suggested that the jittery squiggles observed in electroencephalographic (EEG) recordings of brain activity—including characteristic gamma and delta waves which are interpreted as markers of consciousness—are themselves produced by the microtubule resonances.[50]

Not surprisingly, such a provocative proposal as Orch OR quickly drew strong criticism,[51] and its authors continue to rebut those criticisms.[52] The primary challenge to Orch OR had been the perception that quantum processes cannot operate in the intracellular fluidic environment at body temperatures. However, quantum-based mechanisms have since been shown to play a role in plant photosynthesis,[53] bird brain navigation,[54] our sense of smell,[55] and enzyme function involving long-range electron transfer[56] and proton transfer[57] reactions, and also in microtubules.[58] In fact, according to Hameroff and Penrose: "The evidence now clearly supports Orch

---

50. Hameroff and Penrose, "Consciousness in the Universe," 39–78.

51. Baars and Edelman, "Consciousness," 285–94; Reimers et al., "Revised Penrose-Hameroff," 101–3; Jumper and Scholes, "Life—Warm, Wet and Noisy?," 85–86; also see Al-Khalili and McFadden, *Life on the Edge*.

52. Hameroff and Penrose, "Reply to Criticism," 104–12; Hameroff and Penrose, "Consciousness," 1–17, 39–78.

53. Engel et al., "Evidence for Wavelike Energy Transfer," 782–86; Collini et al., "Coherently Wired Light-Harvesting," 644–47; Panitchayangkoon, "Long-Lived Quantum Coherence," 12766–70; Turner, "Quantitative Investigations," 4857–74.

54. Rodgers and Hore, "Chemical Magnetoreception," 353–60.

55. Turin, "Spectroscopic Mechanism," 773–91; Brookes et al., "Could Humans Recognize Odor," 038101.

56. Gray and Winkler, "Electron Tunneling," 341–72.

57. Nagel and Klinman, "Tunneling and Dynamics," 3095–118.

58. Sahu et al., "Atomic Water Channel," 141–48; Sahu et al., "Multi-level Memory-Switching," 123701; Note: a point-by-point rebuttal to a detailed critique of this proposal can be found in Hameroff and Penrose, "Reply to Criticism of the 'Orch OR Qubit,'" 104–12.

OR. . . . Our new paper updates the evidence, clarifies Orch OR quantum bits, or 'qubits,' as helical pathways in microtubule lattices, rebuts critics, and reviews 20 testable predictions of Orch OR published in 1998—of these, six are confirmed and none refuted."[59]

A Physicalist's Explanation for Spirituality

We see, then, that it is possible to attribute many of the faculties of our mind, personality, and even consciousness itself to the activities of whole brain regions, individual neurons, and even molecules. Science has also shown us that the same can be said about our spiritual experiences. Many have investigated the neurobiology underlying religious practices and spiritual experiences: PET and MRI scans have identified a number of brain regions which are activated during deep meditation,[60] and speaking in tongues.[61] The Apostle Paul attributed the latter to the spirit and completely divorced it from the mind (1 Cor 14:14); however, he would have been working with the science of his own era, much like earlier church leaders did when they were standing behind a three-tiered cosmological model based solely on Scripture (see pages 68–69 above). What should we do in the face of these new scientific observations?

As already explained in the section above, neurons function by converting chemical signals from neighboring neurons into electrical signals. Many of those chemical signals are detected by individual molecules called receptors. A receptor acts like a lock which is opened only by a certain chemical "key" (neurotransmitters). One of the many types of keys is a chemical called serotonin, and there are several types of serotonin receptors which are unlocked by that chemical. Because of the particular distributions of serotonin receptors in different regions of the human brain, certain hallucinogenic drugs can profoundly alter cognitive function by simply activating one particular subtype of serotonin receptor. The purified active ingredient of "magic mushrooms"—psilocybin—is a relatively simple molecule that is metabolized to psilocin which in turn acts through type 2A serotonin receptors ($5HT_{2A}$-receptor) located in the prefrontal cortex and

---

59. Elsevier, "Discovery of Quantum Vibrations"; also see Hameroff and Penrose, "Consciousness in the Universe," 39–78.

60. Shults, *Reforming Anthropological Theology*, 180; Herzog, "Changed Pattern," 182–87; Lou, "^{15}O-H2O PET Study," 98–105; Kjaer et al., "Increased Dopamine Tone," 255–59; Lazar et al., "Functional Brain Mapping, 1581–85; Newberg, "Measurement of Regional Cerebral Blood Flow," 113–22; Newberg, "Neural Basis," 282–91; Beauregard and Paquette, "Neural Correlates," 186–90; Beauregard and Paquette, "EEG Activity," 1–4.

61. Newberg, "Neuroscientific Study," 215.

in limbic regions such as the hippocampus.[62] Careful neuroanatomical and neurofunctional studies show that such hallucinogenic chemicals which stimulate $5HT_{2A}$-receptors (psilocin; LSD; mescaline; N,N-dimethyltryptamine) affect brain areas which are involved with processing sensory information, emotional information, vigilance/alertness levels, and information flow between various brain regions.[63]

More to the point here, users of such hallucinogenic drugs describe experiences which have a very distinct spiritual quality. For this reason, studies have been done to see how those drugs either mimic or modify the spiritual experience. In one seminal study which is much cited in the scientific literature, twenty theological students at Boston University's Marsh Chapel were divided into two groups.[64] Each group participated in a Good Friday worship service held in two different rooms of the building, and were given either active psychedelic drug or a relatively inactive substitute called a "placebo." This study design is referred to as a randomized, placebo-controlled study, and is often used to investigate the physiological or clinical effects of a drug. The placebo is intended to reproduce most of the experience of taking a drug (such as: being closely supervised by a medically trained authority figure; putting a pill in the mouth and experiencing a strange taste; the administration of various test procedures) but must lack all of the pharmacological or drug-related actions of the active drug it is being compared to. The active drug in this study was psilocybin (O-phosphoryl-4-hydroxy-N,N-dimethyltryptamine): the placebo was nicotinic acid, which would produce physiological effects such as pupil dilation and dryness of mouth, but is not known to induce mystical experiences. Those receiving psilocybin reported exceptionally powerful and personally meaningful experiences during that session, some quite positive (peaceful; joyous; euphoric) and others quite negative (fear; terror). Altogether, though, the overall experience was markedly more positive and meaningful than that of the control subjects. When re-interviewed twenty years later, that overall subjective assessment still held true: for many of them, the experience had "increase[d] their depth of faith, increase[d] their appreciation of eternal life, deepen[ed] their sense of the meaning of Christ, and heighten[ed] their sense of joy and beauty." For them, this had been a truly life-changing experience. A

---

62. Pahnke, *Drugs and Mysticism*; Griffiths et al., "Psilocybin Can Occasion," 268–92; Nichols, "Psychedelics," 264–355; Griffiths et al., "Mystical-Type Experiences," 1–12; Griffiths et al., "Psilocybin Occasioned," 649–65; Doblin, "Pahnke's 'Good Friday Experiment,'" 1–28.

63. Nichols, "Psychedelics," 299.

64. Doblin, "Pahnke's 'Good Friday Experiment,'" 1–28; also see Pahnke, *Drugs and Mysticism*.

similar study was later performed using thirty-six volunteers who already regularly practiced some form of religious or spiritual activity.[65] Two-thirds of those who received psilocybin found it to produce either the single most meaningful experience of their lives or among the top five most meaningful experiences in their lives, and they continued to hold that feeling fourteen months later in a follow-up interview.[66]

It is truly intriguing that such a transformative and "spiritual" experience can be had by simply stimulating the appropriate neural pathways using a single molecule acting on a very specific subtype of neural receptor ($5HT_{2A}$-receptors). All of that is happening completely inside of our heads. It is likely that something similar is going on in our internally generated dreams, which can seem so real to us even when we are vaguely aware at the time that "this is only a dream" and much of what is going on is unexplainably bizarre. But we are now learning that something similar can be reproduced when we are fully conscious and awake and not influenced by drugs. Our computers and software are becoming so powerful that one can become completely immersed in a virtual reality that feels so very real. Clayton describes one example in which

> the participant dons a head helmet and a special suit and enters into what seems to be a clearing that surrounds him. He is able to rise in the clearing by breathing in and to lower himself by breathing out; movements are made by gentle leanings from side to side. In the world through which he now "floats" the edges of objects are unclear, and he is able to move through, above, and below the trees and plants at will. This new set of structural couplings between "mind" and "world" can have a profound effect on participants. . . . Some persons emerge from fifteen minutes in this virtual world deeply touched, and (by their reports) sometimes profoundly transformed. Some report that they later experience being embodied in the nonvirtual world in a different way than in the real world, and a few have said, "I am no longer afraid of dying." . . . What occurs when one is in a 360° surround-sound virtual world? It seems clear that the experience gives to subjects a new sense of being embodied—they actually *are* embodied in a different way, thanks to the computer interface. This is why some describe the experience, even months later, as being "within me."[67]

65. Griffiths et al., "Psilocybin Can Mediate," 268–92; Griffiths et al., "Mystical-Type Experiences," 1–12; Griffiths et al., "Psilocybin Occasioned," 649–65.

66. Nichols, "Psychedelics," 271–72.

67. Clayton, "Neuroscience," 195–96.

The Canadian neuroscientist Michael Persinger and inventor Stanley Koren have created an apparatus which has been dubbed the "God Helmet." This device places electronic solenoids over both temporal lobes of the brain (located at left and right sides of the head) and generates complex magnetic fields which have been shaped by signals derived from recordings of a patient during an epileptic seizure which is focused in the hippocampus (based upon work done by other researchers who have drawn a link between epileptic seizures and religious experiences). The intent is that these magnetic fields will stimulate or modulate electrical activity within brain regions of healthy volunteer subjects (perhaps in part by altering the configurations of the tubulin skeletons?), since electricity and magnetism are directly interrelated, being part of a continuum on the electromagnetic spectrum. More to the point, those brain regions which are to be modulated in the test subjects will be the very same ones that produced the epileptic seizure in the original patient. The investigators claim that when the right temporal lobe is stimulated by the device, the subjects sense the presence of a being which they variously interpret as the spirit of a deceased loved one, as an angel or as God himself.[68]

It should be pointed out, though, that others have attempted to replicate this apparatus and the experiments, with the help of these investigators, but now done under double-blind conditions: they failed to reproduce the findings and concluded that the apparatus may enhance suggestibility within the subjects, who were then subtly prompted by expectations (either their own, or in some way that of the investigators) and/or who were previously more religiously inclined and merely attributed the induced-disequilibrium to a Presence.[69] Irrespective of whether the mystical experience is produced by the device or by the inherent suggestibility/religiosity of the subjects, the results nonetheless point toward the human brain itself being able to produce internally some sense of the Divine.

Not only can we attribute certain aspects of our personality, spirituality, and consciousness to the actions of neurons, and even to individual molecules in those neurons, we can also find evidence for influences exerted by our genes. Sociologists can measure religiosity or spirituality—the inclination toward religious thinking, divine mindfulness, and spiritual beliefs—as a psychological trait,[70] and show not only that this is stronger in

68. Persinger et al., "Electromagnetic Induction," 808–30; Persinger, "Religious and Mystical," 1255–62; Persinger, "Vectorial Cerebral Hemisphericity," 915–30.

69. Granqvist et al., "Sensed Presence," 1–6; Granqvist and Larsson, "Contribution of Religiousness," 319–27.

70. De Jager Meezenbroek et al., "Measuring Spirituality," 336–54; MacDonald, "Spirituality: Description," 153–97, 2000; MacDonald, "Spirituality as a Scientific

certain individuals than in others, but seems to exhibit a degree of familial inheritance beyond simply cultural indoctrination (the classical "nature-versus-nurture" question).[71] In fact, one geneticist claims to have linked this trait to a specific gene, although this has not yet been substantiated in peer-reviewed scientific literature.[72] It is hard to understand how something as complex as a behavioral trait like religiosity could be explained by a simple point mutation in a single gene. However, the reader should not think that individual genes code for a specific behavior in the sense that they may otherwise code for a very specific protein: instead, the individual gene functions within a broad network of genes, and its expression influences the range of probabilities of traits or behaviors developing. Hamer's book received widespread attention in the popular press (at the time, it was the cover story for *Time* magazine, and was reported in innumerable large-circulation newspapers around the world) and in public media. However, as of the beginning of 2019, Hamer's scientific data have not yet undergone peer review and publication in the scientific literature, despite his book being published in 2004. However, a genetic study was recently done which revealed a strong correlation between certain genetic point mutations (referred to as single nucleotide polymorphisms) and psychological scores for spirituality or religiosity.[73] More to the point, those mutations were located within the genes for $5HT_{2A}$-receptors, dopamine-receptors, oxytoxin-receptors, and a molecule referred to as VMAT1. The first of these four we have already encountered above in our discussion of hallucinogens which reproduce a spiritual experience (pages 103–104). Dopamine is a neurotransmitter which plays a key role in pleasure and reward, and is upregulated during spiritual experiences such as meditation as well as hallucinogenic experiences.[74] VMAT1 is a molecule which regulates the movement of neurotransmitters such as serotonin and dopamine. Oxytocin is a neurotransmitter which plays a key role in social bonding (usually in the context of sexual reproduction and childbirth, but also in familial and group relationship building).

Indeed, several lines of scientific investigation suggest it is not only humans who have sensed something beyond our material existence. Evidence of ritual burials among Neanderthals lead paleontologists to infer the

---

Construct," e0117701; Eaves, "Genetic and Social Influences," 102–22; Truett et al., "Model System," 35–49.

71. Eaves, "Genetic and Social," 104–8.
72. Hamer, *God Gene*.
73. Anderson et al., "Genetic Correlates," 43–63.
74. Kjaer, "Increased Dopamine Tone," 255–59.

latter also had some form of belief in the afterlife.[75] Likewise, primatologists have documented behavior among chimpanzees which they interpret as evidence of ritual, a precursor of religion.[76]

Non-reductive Physicalism and the Soul

Advocates of non-reductive physicalism fully acknowledge that neuroscience, neuropsychology and clinical neurology are increasingly finding ways to explain every aspect of human behavior, cognition, and the human experience in purely biological/scientific ways, and wonder if the concepts of "personhood" and "the soul" can forever evade the same, especially with the jaw-dropping advances of artificial intelligence. "Research in neuroscience raises the possibility that the concept of a separate, immaterial soul is unnecessary with respect to understanding human life and experience. Unless the soul can be shown to have a separate and identifiable realm of agency, the concept adds nothing critical to understanding humanness."[77] As such, Occam's razor encourages some to discard the immaterial in any model of human ontology. However, other scholars in this area of research are not willing to take that step, fearing that we otherwise reduce ourselves to mere physiological robots or zombies,[78] and propose a thoroughly monist explanation of mind which also accounts for the soul: "Non-reductive physicalism . . . is a view of human nature that is compatible with the neurobiological sciences and potentially more resonant with theology."[79] Before presenting this idea, though, I must first lay down two foundational principles.

First, neuroscience increasingly informs us that cognitive behaviors involve multiple, iterative, and interactive "action-feedback-evaluation-action" loops. For example, in reaching for the cup of coffee on the table, my motor cortex triggers a series of commands to the muscles in my arm and hands to commence the *action*—my eyes and sensors in my muscles send *feedback* to my brain about where my hand is in space and time—a premotor area of my brain *evaluates* where my hand was, is, and should be—and then a new command is issued to the muscles to perform a revised *action*. Most outputs from the brain involve multiple layers of loops like this one, building upon and modifying each other. The more of them working together, the more complicated and refined the output will be. Even our

---

75. Livingstone, *Adam's Ancestors*; Pääbo, *Neanderthal Man*.
76. Kühl, "Chimpanzee," 22219.
77. Brown, "Neurobiological Embodiment," 59.
78. Brown, "Neurobiological Embodiment," 60; Murphy, "Human Nature," 24–25.
79. Brown, "Neurobiological Embodiment," 66–67.

emotional responses can undergo various iterations of action-feedback-evaluation-action loops: as one develops more and more of these loops, one develops what is referred to as emotional-intelligence or maturity. This is also exactly what we see in artificial intelligence and robotics as well: the earliest forms of this technology were really quite clunky, but they became more complicated and refined as more and more lines of computer code and sub-routines were added, until now we see ourselves having conversations with Siri and Alexa.

Some of these processing loops are "bottom-up" inputs, while others are "top-down." Visual-processing provides a good example of both types of inputs. What you eventually "see" is not what is projected onto your retina. The image is indeed captured by the retina's photoreceptors and sent to the visual processing center of your brain, but only after some signal processing has already been done by the photoreceptors, their support cells, and the neurons which carry that partially interpreted information from the "bottom up" to the visual cortex: they have already begun to detect movement, identify edges, enhance contrast, and many other aspects of signal processing.

Other processing units in the brain send information to the visual cortex from the "top-down." Some are dedicated to facial recognition, or associating the image you see with emotional content (informing you whether you should feel happy at seeing that face, or frightened, or surprised). Some add estimations of position, distance, and even lighting. An "illuminating" example of the latter is the image which, a couple years ago, commandeered social media for days with the question "Is this dress black-and-blue or white-and-gold?" These are two completely different color schemes, but one might interpret it as one or the other depending on whether the top-down input to one's visual cortex assumed that the scene being viewed was brightly and directly illuminated, or indirectly and dimly lit. On the other hand, some top-down inputs instruct the visual cortex to essentially throw away portions of the information and accentuate other portions: while scanning a crowd to find your own lost child's face or the coat they were wearing, you completely do not register the other people or unusual things happening within the scene, and then suddenly your child's face seems to leap out from the crowd. A particularly excellent example of one's eyes and brain not "seeing" the same thing is a video, often shown in introductory psychology classes, of a group of people gathered in a circle, half of them wearing black shirts and half of them white shirts, passing two basket balls back and forth to each other, and one is asked to count the number of ball-passes. If the reader has never seen this video, I highly recommend readers do so:[80] following the instructions to the letter will likely lead to a very startling surprise.

Visual perception is one example among many of complex neural networks in our brains: we will now explore how at least one of these might account for the human soul. Warren S. Brown begins his monist proposal of how to explain the human "soul" by providing his working definition of the properties and limits of the human soul: "Much of the popular and theological understanding of the essential properties of humanness and personhood has been tied to the concept of soul," the latter being generally credited with the faculties of thought, volition, emotion, "consciousness of self, personal agency and responsibility, ability to give and receive love, communication with God, the experience of transcendence,"[81] as well as "the property of having deep interpersonal feelings and subjective experiences."[82] He presents a number of examples from his clinical practice in which certain groups of patients with specific forms of brain pathology experienced marked dysfunction in those particular faculties. He also asserts that "most, if not all, of the critical properties that have been subsumed within the Judeo-Christian concept of a soul can be captured in the concept of *personal relatedness*, particularly if one admits the possibility of relatedness to God."[83] He warns his readers that he will substitute the word "soul" with "soulishness" in order to continuously remind the reader that he is attempting to provide an explanation that does not end up turning its object into another thing, or substance, or "essence," and thereby becoming another form of dualism.

Brown then builds his proposal in three steps. First, complex physical movements, intellectual thought, and even emotional responses are products of multilayered action-feedback-evaluation-action loops (as already explained and exemplified above). Second, various forms of intelligence—logic, reason, wisdom, memory, intuition, social intelligence, emotional intelligence, and so on—can thus be seen as epiphenomena emerging from the complicated circuitry, feedback loops and processing units of the human brain. Given those two points, might humanness, soulishness, and personhood likewise emerge from collections of loops and processing units which are dedicated toward the central human tendency toward relatedness? As Brown puts it:

80. One URL which is currently functional is https://www.youtube.com/watch?v=Ahg6qcg0ay4.

81. Brown, "Cognitive Contributions," 100.

82. Brown, "Neurobiological Embodiment," 59.

83. Brown, "Neurobiological Embodiment," 68; also see Brown, "Cognitive Contributions, 101–3, 125.

Thus, soulishness, as described herein, is the product of a progression of emerging capacities. Higher human cognitive capacities (such as language, a theory of mind, episodic memory, self-consciousness, imagination of the future, creativity, and complex problem-solving) emerge out of an evolutionarily expanding neurobiological system interacting with a developing human culture. Personal relatedness is an emergent property of the interaction of these critical human cognitive capacities as they are used interpersonally. Finally, soulishness is a quality of being that emerges from the deepest experiences of personal relatedness.[84]

Brown here redefines concepts which were previously seen as simply *markers* of the human soul now as *cognitive tools* which aid and abet our faculty of dealing with relatedness, and ties all of that together with the concept of the *imago Dei* (our being created in the image of God) which the ancient Hebrews associated with community and relationships (between male and female, with other humans and creatures, and with God), while the classical Greeks and Christians associated it with particular cognitive capacities (language, reason, volition, emotion, memory, and so on). In so doing, he provides a monistic physicalist definition of the human soul as another emergent property of the human brain. It "is a capacity for a particular realm of experience rather than a nonphysical essence inhabiting the body . . . the capacity for, and experience of, personal relatedness . . . emerg[ing] from the operation of an interactive web of cognitive abilities, each of which are present in lower primates, but markedly more developed in humans."[85]

This model of non-reductive physicalism also provides an explanation for the perennial problem that vexed dualist explanations since the days of Descartes and even Augustine: the problem of how the immaterial soul can influence the material body. Brown claims that the dynamic immaterial patterns of consciousness emerging from the various feedback loops and processing units of the functioning biological brain exert "top-down" influences on the material human body. As proof that the latter can be so profoundly influenced by a purely immaterial thought, Brown points to the powerful effect which medical science regularly finds in the medicinal value of the placebo effect:

> One particularly clear and oft-replicated example of top-down causation from the "mental" to the "physical" is the placebo

---

84. Brown, "Neurobiological Embodiment," 74; also see Brown, "Cognitive Contributions," 125.
85. Brown, "Reconciling Scientific and Biblical Portraits," 221.

effect. The placebo effect is created when a higher cognitive process of understanding and belief regarding the effectiveness of a sugar pill (for example) creates a top-down (cognitive process to cellular systems) effect that has been repeatedly shown to enhance the activity of immune cells.[86]

Brown's version of non-reductive physicalism—rooted in the concept that humans have a highly developed faculty for relatedness (toward each other, to themselves, and to God)—has a number of very important theological ramifications.

Given that in one sense relatedness is dependent upon one's ability to *do* the relating to others, does this challenge the soulishness of those who have not yet developed the ability to relate (infants) or who suffer from conditions which erode their ability to relate (brain damage; Alzheimer's disease; senility; autism)? Brown answers in the negative, and holds up "a robust ethic . . . [in which] each individual and community carries the ethical responsibility for the soulishness of all persons within its relational networks, and particularly for those of lesser capacity."[87] And so relatedness is not only dependent upon the one *doing* the relating, but also upon the magnanimity of the others reaching out to relate to the one. Even infants, cognitively impaired, and socially withdrawn individuals partake in relatedness by virtue of the fact that they are valued and related to by others (parents; relatives; friends; society; God).

This noble ethic then raises a few other interesting questions. Amoeba are completely unable to understand us, let alone relate to us. Their chemical and physical sensors are tuned for a variety of stimuli, none of which endow them with the ability to appreciate the essence of who we humans are: they may feel pressure waves in the water that we stir up, or sense chemicals from our skins, but they have absolutely no capacity to put those inputs together in a way that allows them to comprehend *us* at our cognitive level. They are completely dependent (in an unconscious sense) upon us finding a way to interact with them at their level. Christians would certainly assert that God has done exactly that with us humans: but we ourselves have also done so with various other animal species. At the higher level of that relationship-building, many enjoy the relationship they have with dogs, and therefore train them to understand us (obey commands) and communicate with us (sniff out and identify drugs or bombs; alert us to danger or prey) and relate to us (just jump into our laps and "love" us). We've also trained primates to

---

86. Brown, "Neurobiological Embodiment," 66.

87. Brown, "Neurobiological Embodiment," 71; also see Brown, "Cognitive Contributions," 123–25.

communicate with and relate to us: Koko the gorilla and Kanzi the bonobo, who learned to communicate in very emotive ways with their human handlers. That was a result of our reaching out to them, and of their ability to learn how to relate to us. In so doing, we've enhanced their humanness and personhood in our eyes. That experience has also prompted us to ask in a new way, "Do animals also have a soul?" And if yes, what about Siri, Alexa, and Google: as "they" become more advanced and increase their already impressive abilities to relate to us, will they too gain a soul in the process? As Brown wrote: "Whether or not God extends, or would extend, the same form of relatedness to linguistically competent animals or humanoid robots is a question I will leave for theologians and philosophers to debate."[88]

We have seen this asymmetry of interactions between conscious entities before: YHWH breathing into a lump of clay, which in turn *became* a living spirit. Our own soulishness is a result of God wanting a relationship with humans, teaching us to look for him, to understand him, to aspire to his goals? And in the process, he creates his image in us.

This is a substantial improvement over previous explanations for human soulishness (again, not wanting to use the word "soul" lest I risk turning this into a dualistic explanation of a *thing*). Nonetheless, the reader may still prefer to not attribute "the soul" to an epiphenomenon of feedback loops and complex brain circuitry, and may prefer to choose instead to believe in a completely separate immaterial "substance." But that will only distance them from their fellow believers who place just as much value on the Book of God's Works (science) as on the Book of God's Word (the Bible, which never provides an explanation for the soul). They will also continue to alienate themselves from the nonbelievers who cannot embrace a simple dogmatic belief in something that cannot be seen or touched (experimented upon), and will still be left with the intractable problem of how the immaterial soul/spirit influences the material body. In so doing, might they be repeating the same kind of "mistake" that Goheen and Bartholomew identified in the church's handling of the Galileo affair (see pages 69–70)?

Challenges Raised against Non-reductive Physicalism

Given the longstanding tradition within the church of a dualist theology, many arguments have been raised against purely naturalistic or physicalist explanations for consciousness, mind, and soul.

88. Brown, "Neurobiological Embodiment," 71.

It Is Anti-Biblical and Antagonistic to Theistic Belief

As already noted above, a majority of Christians (and possibly all non-Christians) seem to think that substance dualism in some form is central and essential to Christian theology. It is not. A large number of biblical scholars assert that the ancient Hebrew and New Testament writers were certainly holistic in their theology, but nonetheless occasionally employed wording which seemed quite monist. We have shown above that the Bible does not explicitly teach Cartesian dualism (although it is easy to interpret certain biblical passages anachronistically from a dualist perspective).

Misplaced View of Authority

Some resist it on the grounds that the primary reason for even considering non-reductive physicalism is simply that it more satisfyingly answers new challenges raised by modern science: in other words, that the proponents of physicalist explanations are elevating the authority of science above the authority of Scripture (even though, once again, Scripture does *not* explicitly teach dualism). This unnecessarily emotionalizes the matter, and draws lines where none need to be found. This is the same reason that much of the church currently resists any explanations which refer to biological evolution, especially the evolution of humans from a hominid ancestor which we share in common with the chimpanzees and Neanderthals; yet an increasing proportion of Christians—scholars and laypeople alike—are conceding that the evidence for biological evolution is becoming overwhelmingly convincing. This same reason for resisting change was also raised when Copernicus and Galileo challenged the traditional church model of the cosmos (see pages 69–70 above), and yet the church eventually accepted that the scientific evidence was decidedly against the traditional theological model and in favor of the scientific Copernican model. A large fraction of Christians have now come to accept that the universe came into existence through the big bang (and they reinterpret the first chapter of Genesis accordingly). I doubt that any Christians will dismiss weather reports issued by the Weather Office simply because those reports are founded upon scientific models which work solely upon gradients of temperature, air pressure, and humidity, rather than the divine mechanisms clearly delineated in the book of Job (chs. 37 and 38). No Christians today will attribute their emotions or volition to their kidneys, intestines, or the organ we call the heart. All of these are examples of modern day readers setting aside the explanations of prescientific biblical authors writing from the point of view of the science of their own day, and allowing the "authority" of modern science to guide an appropriate change in contemporary theology. More importantly, we have grown accustomed to accepting the authority of science in all those other cases: why not also in this area of human ontology?

It Challenges the Concept of an Eternal Existence

There is a fear that if the mind and soul are "simply" emergent properties of the human brain, then when the brain dies, the personhood ceases to exist. In other words, that we can no longer look forward to immortality. There is no justification for this fear for those who believe in the concept of resurrection, which involves a reembodiment of the soul (death/reembodiment are much like sleep/waking: someone falling asleep does not normally need to fear waking up again when all the right biological conditions are restored). This fear is rooted in one's understanding of continuity: that one's essence must continue to exist beyond death of the physical/earthly body, and must continue in some form of new/heavenly body. An eighty-year-old can look back at pictures of themselves at eight and twenty-eight years old and still identify with the person that they see. But the body that the eight-year-old had is not the one that the twenty-eight-year-old had, and both those bodies are no longer the one that she now has. Do we really identify with our body? (Given the amount of attention some of us pay to our bodies, one might possibly say, half facetiously, that the answer is a resounding "yes!") Each of us has a truly unique body: literally, one-of-a-kind. Outside of identical twins, it would be rare (impossible?) to find two people with exactly the same face and set of fingerprints. There are also unique arrangements of anatomical features. A comparison of the blood vessels on the arms, back of the hands, or the legs will show that these vary between individuals. The same can be said for the branching of the airways in our lungs, the innervation, musculature, the glomeruli in our kidneys, the placement of hair follicles all over our body, and many other parts of our anatomy. The molecular details of the cells in our blood vary, and the same can be said about the molecular details of the cells comprising our organs, necessitating a blood/tissue type matching before a transfusion or transplant can be done. Even identical twins will differ markedly in two fundamental respects: the collection of cells and molecules which make up their immune system, and the interconnections of the synapses in their brain. The former continually evolves over the course of their lifetimes as they are exposed to various antigens, viruses, and bacteria at different times and places in their unique life histories, and the latter evolve as the twins learn new skills and acquire unique memories over the course of a lifetime. In these and so many other ways, it is conceivable that each one of us is structurally unique to every other human body that has ever existed in all of human history.

And yet, the unique body which I have today is not, to be precise, the same one I had ten years ago, and will not be the same one I possess in another ten years. I am not referring simply to the fact that our bodies change as we proceed through life stages (infancy, puberty, maturity, and senescence), gain/lose weight, and suffer injuries leading to scars and amputations. In addition to those changes, we are constantly breaking down every structure in our body and exchanging it with new molecules placed in more or less the same location and configuration. The dust that collects in my office was not too long ago the outermost layer of my skin, which has since been replaced with new cells generated by deeper epidermal layers. Other molecules which were once part of my pancreas or eyeball have long ago been excreted, having been replaced by newly made proteins: the yellow color of urine itself owes to break-down products from our red blood cells, which circulate on average for only 100–120 days before they are replaced. There is a constant discarding and replacing of molecular material, most (all?) of which has at one point in time been part of the bodies of many other individuals throughout history. In this sense, the "Transporter" of the sci-fi series *Star Trek*—the machine they would use to dematerialize their bodies and send themselves to far distant locations where they were rematerialized—is not science fiction: our bodies have in fact been in a continuous process of dematerialization and rematerialization throughout our whole lives!

To return to the subject at hand: though we might have a strong emotional attachment to a physical body which has been constantly changing over our lifetime, do we *really* identify with it? Or do we identify with the continuity of our memories and life experiences? Many philosophers conclude that the latter is in fact the case.[89] When we reach the age of eighty years old, we do not identify as much with the body we had when we were eight as we do with the experiences we were living at that time. When that eighty-year-old remembers their life at twenty-eight, they are probably less likely to say "I've still got that body" as to say "I'm still that stubborn, overconfident kid who was scared but just trying to be brave and look cool." Many books and movies have playfully worked with the premise that one could swap bodies with another person or even an animal: in each case, the identity of the character follows the memories and experiences of that

---

89. Murphy, "Nonreductive Physicalism: Philosophical Challenges," 103–8; Williams, "The Self," 46–63; Green, *Body, Soul*, 142, 178.

A SYNTHESIS OF MODERN THINKING ON OUR INNER BEING   117

individual, not the body that they once occupied. Other books and movies—and real life stories—revolve around the psychological trauma which accompanies a complete loss of all memory, including one's name and life experiences. Other stories hold up the nobility of dying but being long-remembered (even if that death was seen as premature), versus the poverty of living a long life but dying lonely and immediately forgotten. If one were ever in a position of being able to choose between two future technologies, one which could forever halt the degeneration and death of our body but which would immediately erase all memories and life experiences, or another which could maintain the collection of those memories and experiences in perpetuity but would destroy the body, it is probably safe to say that most people would choose the second technology over the first (although some would likely prefer to decline any changes at all and simply face a natural death). And if those technologies were improved such that one could keep their original body for eternity but with someone else's memories and experience transplanted in, or have their own memories and experiences transplanted into a fully functioning donor body, many more would choose the second option. So, it is actually the continuity of one's personal memories and life experiences which is most important, and non-reductive physicalism can still account for that form of continuity. We have also seen above that who we are—our soul—is also defined by our relationships with others: with family, friends, and God; that too can continue beyond physical death and can be reconstituted in some other kind of body. Thus, there is no need for the fear that "if Cartesians can no longer live in their disembodied souls, then the souls of Christians can no longer be immortal. This is an error because the concept of a disembodied soul, mortal or immortal, is not fundamental to Christian eschatology. Resurrection of the body is."[90] "The most important development in the New Testament is the belief that the essential personality (whether called the *psychē* or the *pneuma*) survives bodily death. This soul or spirit may be temporarily disembodied, but it is not complete without the body, and its continued existence after bodily death is dependent upon God rather than a natural endowment of the soul."[91] The second half of this book will focus on the afterlife, resurrection, and immortality.

90. Peters, "Resurrection," 305; also see Peacocke, "Christian Materialism?," 146–68.
91. Murphy, introduction to *Neuroscience and the Person*, vi; also see Murphy, "Human Nature," 20; also see Murphy, "Nonreductive Physicalism: Philosophical Challenges," 97–98.

It Diminishes the Sovereignty and Grandeur of God

Some Christians—particularly those having a more traditional understanding of biblical theology and/or encountering these thoroughly materialistic explanations for consciousness, mind, and soul for the first time—may feel that these diminish one's view of God or of his creative acts, and lessens Christian faith. This is understandable, *but completely unnecessary.* One can look at stunning pictures of galaxies, star clusters, solar systems, and planets, and be amazed at how these are a result of mathematically defined physical forces such as gravity and quantum mechanics, and still believe that ultimately God was behind all that. The same can be said for breathtaking panoramas of mountain ranges, the Grand Canyon, oceans, and lakes—all formed by "simple" mathematically defined geologic processes like gravity, erosion, tectonic activity, and thermodynamics—and still be amazed at God's creative sovereignty. Likewise, God could ordain and use ecological and biological processes to craft a Brazilian rainforest eco-system . . . a few "simple" atmospheric rules based on thermodynamics to give us the wide range of weather patterns . . . as well as biochemistry and genetics to orchestrate the knitting together of a human baby. We have completely naturalistic explanations for earthquakes, for how the heart beats, for vision, for photosynthesis, *in vitro* fertilization, and many other processes which were totally inexplicable even one hundred years ago: yet none of those explanations need to lessen the awe we feel at God's wisdom, creativity, foreknowledge, nor any other of his attributes. Given all these examples, why not also accept that he could use processes and principles defined by biology, biochemistry, physics, and quantum mechanics to build up the complex neural networks that give us consciousness, mind, personality, and soul? Does this make him any less of an all-powerful God?

Free Will and Creativity

There is also a philosophical difficulty within monism: how to account for free will or spontaneous creativity in a way that easily and fully convinces all experts in that area of research. Being purely naturalistic or physicalist in nature, they would seem to be bound by determinism. That is, physicalists hold that any physical event has a physical cause: a thought in the brain triggers a cascade of neural firings which in turn activates the muscle which then moves the arm. But what first triggered the thought itself, which culminated in the moving of the arm? Or the emotion to write a song? Or to suddenly decide to take up hang-gliding? Or motivated Seurat to invent Pointilism in order to paint his subject, or Picasso to take a Cubist approach, or Beethoven to compose his symphonies?

Non-reductive physicalism also faces this challenge. But it does go further (in my estimation) than the others. That is, the dynamic interplay of the neural circuits which produce the complex emergent phenomena of mind, consciousness and soulishness may also lead to unexpected, unpredictable, even chaotic behaviors[92] ("chaos" used here in the sense that mathematicians and physicists mean it, not in the evil or destructive sense that theologians mean it, nor the cataclysmic apocalyptic sense that others might mean it). Neuroscientists have discovered behavioral control mechanisms which lead to spontaneous changes in lower animals: bacteria and fruit flies which exhibit sudden sporadic changes in direction produced by "randomness generators" in their propulsion mechanisms which are believed to increase their odds of success at locating food sources or evading capture. It is entirely possible that such mechanisms are also employed within the circuitry which guides our own behaviors and cognitive thinking, leading to innovation and creativity, or at the very least help us to choose between two equally advantageous decisions. "*Buridan's Ass*" is a medieval philosophical fiction in which the animal starved to death between two equidistant and equivalent piles of hay, being completely unable to make a justifiable choice between the two. We might otherwise suffer a similar fate when trying to decide "What color of shirt to wear to work today?" or "Which door to open when fleeing an approaching ax-murderer?" but we somehow come up with a decisive answer that we cannot explain beyond saying: "I just made a choice." Some bring in the Heisenberg Uncertainty Principle, possibly in the context of the quantum superposition states referred to in the Penrose-Hameroff Orch OR proposal (see page 101).[93] It is also argued that the retrojection of quantum information backward in time in Orch OR can also contribute in part to creativity, spontaneity, and the perception of volition.[94] As such, the manifestation of the soul completely transcends the relative simplicity of the individual processing units, neurons, and synapses in the same way that the organismal manifestation of the flock of starlings transcends the nature of the individual birds of which a murmuration is made up.

---

92. Jeeves, "Brain, Mind, Behavior," 95.

93. Jeeves, "Brain, Mind, Behavior," 94; Schwartz, "Quantum Physics," 1309–27; Hameroff, "How Quantum Brain Biology," 1–17; Korf, "Quantum and Multidimensional," 345–55.

94. Hameroff, "How Quantum Brain Biology," 93.

Non-reductive Physicalism and the Human Spirit

We have seen above how millennia of pondering human existence, much of that fertilized by introspection, meditation, study of Scripture, divine revelation, and scientific investigation have now given us explanations for many aspects of our *seemingly* immaterial being. Mind and consciousness are highly likely to be emergent properties of large numbers of neurons and neural processing units working together, and may also include specific roles for certain molecules and for quantum physical mechanisms. Our personality emerges from that as well, and is shaped by our experiences, cultural upbringing, the knowledge and memories we have acquired, as well as genetic factors and neuroanatomical development (and injury/damage).[95] Something similar could be said about our soulishness which emerges from signal processing in neurons directed toward relatedness, and employing cognitive faculties such as language, volition, theory of mind, mirror neurons, memory, and emotion to create our experience. All of these emergent (and seemingly immaterial) properties can easily direct and influence the actions of the material body. So in all four of these aspects of human immaterial being—consciousness, mind, personality, and soul—Christians and nonbelievers can be "on the same page," and without being in contradiction of any Scripture. There is no longer any need for us to divide on these matters. If we do, such biases are not scientifically based (built on empirical facts), but are largely philosophical and emotional in nature.

But Christians *do* go one step further, being willing to accept that there *is* a God—an immaterial, spiritual being—who desires to be in relationship with humans. Does that require an additional component to our inner being? Several contemporary scholars think not. Murphy suggests that "no special faculty [is] needed in order to experience religious realities. What makes the experience religious is a meaningful combination of ordinary experiences, under circumstances that make it apparent that God is involved in the event in a special way."[96] "I submit that for such an experience, nothing is needed beyond the ordinary neural equipment that we all possess."[97] She aligns her view with that of Maurice Wiles:[98] "Revelation is not the result of special action on God's part, but is to be explained in terms of special sensitivity of some people to God's general action."[99] Brown offers a wonderfully

---

95. Eaves, "Genetic and Social," 102–22.
96. Murphy, "Nonreductive Physicalism: Philosophical Issues," 143, 147.
97. Murphy, "Nonreductive Physicalism: Philosophical Challenges," 102.
98. Wiles, "Religious Authority," 181–94.
99. Murphy, "Nonreductive Physicalism: Philosophical Issues," 147.

poetic definition of this emergent property: "Soul is the music made by an orchestra of cognitive players performing in the context of interpersonal (or intrapersonal) dialogue. Played out in relationship to God, who chooses to be in dialogue with his human creatures, the cognitive capacity for personal relatedness embodies spirituality."[100] All three are saying that humans are equipped with faculties in their soul (and mind, consciousness, and personality) which they routinely employ in their interactions with other people and the rest of creation, but can also develop a sensitivity in those same faculties toward God as well.

Others might choose to believe, however, that we *do* have an additional component: the human spirit, which is different from the soul. The writer of Hebrews refers to the word of God being able to divide soul and spirit (Heb 4:12). How might these two be different? Might it be that whereas the soul orients itself toward *physical* beings (our own self, other humans, and animals) and yet can also be attuned toward spiritual beings (God), the spirit is primarily oriented toward God—it is particularly receptive to an *experience of* God, a *relationship with* God—and can communicate that experience to the soul (which is also in communication with the mind and body)? Being spiritual and immaterial, though, the existence of such an entity is beyond the reach of materialism and science: the very definition of *super*natural. It cannot be tested or studied using scientific tools, nor can it provide scientific explanations of material events. Nonetheless, it does provide a *theological* perspective on Christ's claim that "no one can see the kingdom of God unless they are born again" (John 3:3). Nicodemus—a prominent Jewish leader, highly educated Pharisee, and teacher of the Jewish law puzzling over this statement from a naturalistic angle—stumbled over this truth claim (John 3:4, 9). The first time humans are born, they are endowed with a biological investment—a brain—from which emerges consciousness, a mind, a personality, and a soul. Perhaps Christ is teaching that humans must be born in a different way—"born again"—in order to receive another entirely different asset which opens their mind to a spiritual experience (perhaps the same asset that Genesis says Adam and Eve lost when they "died" after taking the fruit from the tree; Gen 2:17). It seems quite unlikely that the "death" spoken of in that passage of Genesis was in any way related to physical/biological death, which has been present in nature for hundreds of millions of years, long before there were humans. It also seems to have had little or nothing to do with death of the consciousness, mind, personality or soul, since Adam and Eve (and all their descendants right up to today) continued to love, laugh, fear, and

---

100. Brown, "Reconciling Scientific and Biblical Portraits," 221.

otherwise relate to each other and to God (although the latter relationship was/is now markedly strained, to say the least). A semblance of that spiritual experience can be artificially reproduced by hijacking many of the very same neuronal processing units which allow for experiencing that relatedness—as hallucinogenic drugs, carefully applied electromagnetic waves, and even certain virtual reality experiences may do—in the very same way that one can artificially experience an interaction with a loved one or relive an event by electrical stimulation of specific brain regions.

Could Paul be pointing to this when he writes: "In the same way, the Spirit helps us in our weakness. We do not know what we ought to pray for, but the Spirit himself intercedes for us through wordless groans. And he who searches our hearts knows the mind of the Spirit, because the Spirit intercedes for God's people in accordance with the will of God" (Rom 8:26–27)? Paul taught that we should "put off [the] old self, which is being corrupted by its deceitful desires; to be made new in the attitude of your minds; and to put on the new self, created to be like God in true righteousness and holiness" (Eph 4:22–24). Might this be the work of the spirit, influencing the thoughts of our soul and mind, as we immerse the latter in meditation, prayer, dialogue with other believers, and study of Scripture? In the very same way that living in a specific human culture, speaking a certain human language, embracing a certain human historical narrative, and practicing a distinct way of life all shape one's mind, personality, and consciousness?[101] This would be a form or extrapolation of the Moral Influence Theory of Atonement.[102]

Remaining Questions and Concerns for a Christian View of Ontology

Many of the questions which the ancient Hebrews, the early Christian church, and the church fathers wrestled with still linger. But now, in light of the recent developments in scientific and philosophical scholarship, we can now add a whole new set of questions and/or bring to bear a whole new set of observations upon those long-standing questions. But putting that new information on the table will require some adjustment to Christian theology, something which the church has shown a track record for resisting strongly (until it eventually capitulates). The first two questions explored here may be seen to have less immediate practical application to living a

---

101. Eaves, "Genetic and Social," 108–13.
102. Janssen, "'Fallen' and 'Broken,'" 40.

Christian life, but the other questions have much more profound implications in that respect.

Augustine wrestled long over the question of "What is the origin of the soul?" In Augustine's mind, the human soul and human spirit were synonymous concepts. Some readers may agree that we now have an answer to his question: the "soul" is an emergent property of our brain, a panel of cognitive skills which humans have evolved for relating to physical beings (our own self, other humans, and even animals) but now also use in relating to God. But for those readers who might continue to hold a dualist view—that the human soul and/or spirit are independent of the body—then Augustine's question still remains. As we noted previously, Augustine reduced the answer to four possibilities: logic dictates that it either preexists the body (and is either assigned by God, or chooses by itself, to be joined to the body), or it originates with the body at conception (being created anew by God, or arising through generation). Reframing these four logical possibilities from a more philosophical perspective: has our soul always existed and we *discover* it during our life on earth (along the lines of Plato's Forms and Ideas), or is it *progressively formed* during our life on earth (along the lines of Tertullian's view of the infancy, puberty, and maturation of the soul [see page 54], or Scott's and Phinney's view of "developing hominization" [see page 89])? It would be a huge stretch to suggest that Paul leaned toward the preexistence of the human soul on the basis of his comment to Timothy that "we brought nothing into the world, and we can take nothing out of it" (1 Tim 6:7), although Paul clearly believed in the preexistence of Jesus (1 Cor 8:6; 2 Cor 8:9; Phil 2:6–7; Col 1:15–17). Augustine was never able to settle upon any one of the four possibilities that he originally identified. In a letter to Optatus, he wrote: "I have therefore found nothing certain about the origin of the soul in the canonical Scriptures."[103] Aquinas leaned toward the third option, thinking that the soul was created by God a few weeks after conception.[104] Bjork readdressed this same question[105] by comparing traducianism against soul creationism and emergence. (Traducianism is the idea that the body and the soul of each person are propagated from the generation that preceded them, and ultimately are propagated from the body and soul of Adam [similar to house plants being propagated by plant cuttings], while soul creationism refers to God creating a new soul around the time of conception.) Note that those three mutually exclusive possibilities that Bjork compared are akin to Augustine's first, third, and fourth proposals, respectively (see above). In

---

103. Augustine, letter 190 (to Bishop Optatus), in *Letters*, 271–88.
104. Barbour, "Neuroscience," 252.
105. Bjork, "Artificial Intelligence," 95–102.

doing so Bjork, as well as Scott and Phinney[106] point out that viewing the soul as an emergent phenomenon easily answers the special case of identical twins (these can/would have different souls, since the latter emerges as the brain matures), while soul creationism and traducianism do not (these would require the soul to somehow split in half along with the developing embryo). I leave it to the reader to decide how they answer this millennia-old question raised by Augustine.

That very same question can be pondered within the context of biological evolution. Do animals also have a soul? If one sides with Descartes by denying animals of this (or even consciousness),[107] then at what point in our own evolutionary history did we humans acquire it as we evolved away from other animal species? The theory of biological evolution—accepted by many believers and most agnostics/atheists alike—holds that humans evolved from lesser animals through a gradual process stretched out over millennia. Irrespective of how one describes the immaterial aspect of humanity, when did our human lineage first acquire it? How far back in our hominin ancestry would we find a soul/spirit? Can one interpret the receiving of the "breath of life"—a gift given to animals and humans alike (see Gen 1:30; 6:17; 7:15, 22)—as receiving a soul/spirit? Some believers who attempt to harmonize their theology with evolution theory will answer this question by pointing to the *imago Dei* and the "breath of life" passages: that God took existing hominids and made something novel out of them (by metaphorically breathing something new into them).[108] However, we have seen that the "breath of life" passages do not distinguish between animals and humans. Also, suggesting that pre-hominids were "updated" or "remodeled" to produce the first full humans raises other theological and ethical questions. One could question whether this was equitable to the other hominids who were left in spiritual darkness. But more importantly, this remodeling would need to have included genetic changes or social taboos which absolutely precluded reproduction with the non-remodeled hominids all around them: otherwise, there would have been crossbreeding between them, leading to hybrids with varying amounts of humanness in them. (This assumes that inheritance of a soul/spirit proceeds along genetic lines [soul traducianism]; soul creationism and viewing the soul as an emergent phenomenon do not raise these concerns.) We now have abundant genetic evidence that humans crossbred with Neanderthals and Denisovans: in fact, we have a specimen which

---

106. Scott and Phinney, "Relating Body and Soul," 97.

107. Scott and Phinney, "Relating Body and Soul," 93; Swinburne, *Evolution of the Soul*, 183.

108. McGrath, "Soteriology: Adam and the Fall," 252–63.

radiometric dating and genetic analysis tell us came from a human who lived forty thousand years ago and whose ancestors included a Neanderthal only four to six generations prior to the birth of that individual.[109] (As an aside, we also have a specimen of a female who was parented by a Denisovan father and a Neanderthal mother roughly fifty thousand years ago.)[110] What would be the "soul status" of the various hybrids which arose from crossbreeding between "remodeled" hominids and those that were not? In this context, Davis A. Young raised a particularly troubling conundrum:

> Missionary strategists would be put in the very uncomfortable position of identifying those groups of anatomically modern "people" who are not descendants of Adam and Eve and thus not really human. . . . As non-image-bearers, such "peoples" are therefore not sinners and are ineligible for salvation. They do not need it. Missionary activity among such groups is unnecessary. We do not evangelize non-humans.[111]

Hand-in-hand with that question: what exactly is meant when YHWH tells Adam that "in the day that you eat of it, you will die" (Gen 2:17)? As we already noted above, this clearly did not pertain to physical or biological death (of the body), nor did it pertain to death of their mind/personality, nor an end to their consciousness; again, some readers will be able to also conclude this death did not pertain to death of the soul or spirit, since they would see the latter as emergent properties of the brain. So what did God mean when he said "in that day you will surely die" (Gen 2:17) and what precisely did Jesus mean when he said we "must be born again" in order to see the kingdom of God (John 3:3)? The answer to the latter may depend on how one defines "the kingdom of God," something we will explore in the next chapter (page 167).

Other difficulties and questions arise within discussions of Incarnation. What is meant when John writes that Christ breathed the Holy Spirit on the disciples (John 20:22)? Does the Holy Spirit "inhabit" a human body in any way like that in which we provisionally suggest our soul and/or spirit do, especially recognizing the infinitude of God's spirit? This pertains not only to the indwelling of believers in general, but also to the humanity/divinity of Christ specifically. Siemens raises questions regarding the conundrum of the material and immaterial existing *together* (embodied spirits), but also that of immaterial existing *without* the material (disembodied spirits): the latter

109. Fu et al., "Early Modern Human," 216–19.
110. Brown et al., "Identification," 23559; also see Warren, "Mum's a Neanderthal," 417–18.
111. Young, "Antiquity and Unity," 380–96.

would include long-dead individuals (the "Samuel" conjured up for King Saul by the medium at Endor [1 Sam 28:11–14]), angels, demons, and even Satan himself. How does a disembodied personality exist, and how does that personality interact with material humans: it has no brain to generate its thoughts, nor nerves and body parts to direct those thoughts outward, so how does a spirit think and act? This question is essentially the same as that which vexed Descartes and many other philosophers/theologians before him: that is, how does the immaterial aspect of a human exert control and influence upon its own material aspect? For that matter, how do the human soul, mind, and personality exist after death and before resurrection?

Summary and Conclusion

Christian theology is founded firmly upon Greek philosophy and Hebrew theology: both were built up around an internal structure shaped by ancient Sumerian, Akkadian, Egyptian, and Greek mythology.

Ancient Near Eastern thinkers going as far back as several millennia and their scholarly descendants as recently as just a few centuries BCE, including those who authored the Hebrew Scriptures, were relatively unanimous regarding explanations for the origin of the human body. For them, it was fashioned by a divine action upon non-living clay/dirt, often combined with some divine element (tears; blood; saliva; flesh; breath).[112] However, they had relatively little to say about the nature or origin of any putative immaterial aspect of the human experience. Clearly many believed in "spirit," but they left very few clues for us to know how they would describe or explain the latter. The Hebrews among them may have held a holistic view of ontology, but one that clearly leaned toward dualism.

For at least four thousand years, there seemed to be very little development of the concept of mind, soul, or spirit; only variations on a common theme. But then there was an explosion of thinking on these concepts—which brought a whole paradigm change—over the next thousand years: Greek philosophers/theologians (first secular ones, and later Christianized ones) over the course of a millennium stretching from the fifth century BCE to the fifth century CE propelled a dramatic evolution in our explanations and descriptions of that immaterial aspect of human nature: "Anthropogony 2.0" was an unequivocally dualistic perspective. Collectively, they developed a number of elaborate and detailed schema and ideas about the human mind, soul, and spirit, based largely upon arguments, conjecture, and subjective interpretation of ideas and beliefs rather than objective scientific

---

112. Walton, *Lost World of Genesis*, 68.

experimentation. A broad brush summary of Patristic theology regarding anthropogony holds that humans comprise a soul/spirit which is a creation or emanation from God and is immortal, housed within a fleshy body which was also divinely created, albeit from clay, is tainted by sin, and is therefore mortal. Our soul/spirit is destined for reunion with God in heaven, and this transition is facilitated by meditation upon the divine; that transition may also include the body.

Then, especially over the past several centuries, science picked up the ball where it had been left and continued that exploration, bringing yet another entirely new perspective—"Anthropogony 3.0"—which has challenged Christian thinking in its attempts to do away with the immaterial and provide solely naturalistic explanations. This new perspective claimed our bodies were *not* fashioned from inanimate clay in a divine creative act, but have evolved over millions of years from previous generations of animals, the latter in turn arising over billions of years from single-celled organisms. Our minds and personalities, they would say, are merely epiphenomena emerging from an extensively connected network of neurons, and we have tentative naturalist explanations for consciousness. They conclude there is no such thing as "spirit" or "soul," "heaven" or "hell." Christians were forced to choose between either rejecting these assertions outright (which has generally been the choice made by non-academic Christians), or revising our thinking to come into line with theirs but without losing essential theological concepts (generally the choice for Christians within the academic setting). The fruit of this Christian revisionism has included non-reductionist physicalism, which provides a scientifically satisfactory explanation for "the soul" that does not need to conflict with Scripture or Christian theology. Can it be that the soul is an evolutionary adaptation which increasingly nuanced an ability to relate to others (and conferred many advantages with that), but now, in humans, can be employed in our efforts to experience a relationship with our Creator?

But now, philosophers and scientists are rejoining theologians in the belief that we are more than just the material. There is a (seemingly) immaterial component—consciousness, both individual as well as universal—which we just cannot yet explain:

> More and more philosophers and scientists are acknowledging that consciousness is an essential aspect of the universe which must be included in any theory [of philosophy of mind] . . . it is built into the quantum reality of the universe . . . there is a growing consensus that human consciousness (or mind) cannot be explained either by completely reducing it to brain functions

(monism) or by separating it substantially from the body (dualism). The former cannot account for subjectivity, and the latter cannot elucidate the interaction between body and mind.[113]

The dispute between these two poles goes on, and quite a number of recent proposals have been made even in just the past several decades for modern dualism (emergent dualism; holistic dualism; naturalistic dualism; integrative dualism; Thomist dualism; idealism; pluralism; Aristotelian hylomorphism) versus monism (non-reductive physicalism; dual-aspect monism; dipolar monism; reflexive monism; constitutional materialism; emergent monism; deep physicalism; eliminative materialism), as well as several reductive forms.[114] Modern Christian thinking on this subject is still adapting to these proposals and the scholarly discussion. Perhaps we are on the verge of seeing "Anthropology 4.0"? While it may ultimately be not so important to ascertain exactly where/how the human body originated, one's view on that question greatly impacts other bigger issues such as one's conviction regarding determinism versus free will, sin and the fall, as well as reconciliation and atonement.[115]

---

113. Shults, *Reforming Anthropological Theology*, 181–82.

114. Cortez, *Theological Anthropology*, 92; Brown, "Neurobiological Embodiment," 58–76; Green, *Body, Soul*, 29.

115. Janssen, "Fallen and Broken."

CHAPTER 4

The Afterlife

In the preceding chapters, we considered the evolution of Judeo-Christian thinking in particular, as well as that of other societies in general, upon the question of "What is the self?" Much of that discussion revolved around the seemingly immaterial aspects of human ontology. Occasionally, that discussion brushed up against the concept of immortality: many societies have believed or do believe that the mind/soul/spirit continues to exist beyond death. For those believers, two topics which naturally flow from discussions of anthropogony and human ontology are those that deal with the resurrection and the afterlife. Once one has lived a life on earth and stands at death's door so to speak, what is next? This question and others like it have also been asked by philosophers and theologians for the past many millennia. And once again, Judeo-Christian thinking on this question has evolved in response to the changing ideas of the cultures and societies with which we grew up over the past four millennia or more.

A traditional view among scholars has been that the Old Testament described Sheol as the final destiny for humans following death: Sheol was thought to be a gloomy nonexistence where there is no interaction with the living or the dead, or even with YHWH, and from which there is no return. Philip Johnston claims that recent archaeological advances pertaining to the ancient Hebrew understanding of the afterlife compel a reevaluation of the traditional understanding of Sheol: despite many recent discoveries of artifacts in the Middle East and advances in exegesis of Hebrew Scripture, "there has been little attempt to provide a detailed synthesis since that of Martin-Achard, first published in 1956, nearly half a century ago."[1]

I will characterize the evolution of the concept of Sheol, showing it to originally be a final resting place without a postmortem judgment, but later to refer to a state of transition or of condition—the prolongation of

---

1. Johnston, *Shades of Sheol*, 16.

an unfulfilled life characterized by isolation and abandonment—which *can* be reversed. Likewise, its opposite is also not a place, but is instead the fulfillment of rejoining one's ancestors and of being connected to a long line of descendants who remember and honor the individual. Both of these ancient Hebrew views seem to stand in stark contrast to the New Testament understanding of heaven, hell and final judgment. Nonetheless, some posit that those latter ideas were already present in "embryonic form" in the Old Testament texts, and that a growing philosophical frustration with the possibility that the righteous and the wicked shared the same final fate—with no reward or recompense for the life that one lived—energized the emergence of that full-blown Christian theology. Again, though, we will first explore the philosophy of other civilizations which preceded and surrounded the ancient Hebrews, given that Judeo-Christian thinking has long shown a tendency to evolve in tandem with the surrounding zeitgeist.

Prehistoric Views of the Afterlife

Paleontologists have collected evidence of ritual burials dating back well over a hundred thousand years.[2] Certain bodies were carefully laid out in a fetal position or prone as if in sleep, sometimes facing east (possibly anticipating the dawn of a new day?). Some were embedded in flowers and/or red ochre, or were accompanied by food, jewelry, weapons, and even other humans (loved ones? servants or assistants?). These finds and others betray some kind of belief in an afterlife. Although we will likely never reclaim any details of what their understanding of that afterlife looked like (given that those archaeological finds long predate the invention of writing and recorded history), it is clear that humans have contemplated certain aspects of this belief long before the writing of any Scripture just a few thousand years ago.

Observing a dead body is a foreign experience for most modern readers. For the ancients, however, it was commonplace. They would witness the dead bodies slowly sinking down into the soil. This would, within their minds, connect death with an underground existence (if dead bodies always floated up into the sky, or always turned into a smoke or vapor that lifted up, ancient humans would likely connect the afterlife with the sky; if the bodies always liquefied and formed rivulets to the nearest stream or body of water, they might naturally connect the afterlife with the ocean). The inability to locate that underground realm would suggest it would be found very deep, far removed from life and even the source of life (deity). This made sense, since the depths of the earth would be the polar opposite to the place

---

2. Pääbo, *Neanderthal Man*; Livingstone, *Adam's Ancestors*; Davies, *Death*, 24.

where one would naturally expect to find a beneficent deity. Humans would naturally expect the latter to be found "up"—in the sky, the stars, or in the heavens—since, for whatever reason, it seems to be engrained in our thinking that "up" is the preferred place to be. Royalty have often built castles at the tops of mountains; even today, the most expensive homes can be found at heights above the rest of the surrounding city. Temples, throne-rooms, thrones, speaker's platforms, VIP seats at ceremonies, and even preferred box seats at entertainment and sporting events today, all have some form of elevation built into them. High-flying birds, rather than earth-burrowing animals, are often the symbol of omniscience and freedom. Naturally, then, for those ancients, that meant that the place of the dead must be "down," deep under the ground's surface.

Certain ancient ritual burial practices in arid climates (from the desert low-lands of Mesopotamia to the mountaintops of Tibet or Peru or the coastal regions of Chile) would often lead the ancients to associate death with dust. That is, when the body was left in a sealed vault which could be later revisited (such as a natural cave or an artificially constructed pyramid or tomb), the living would return to find the bones and dried remains of the flesh. When the latter were disturbed—often with the intent of gathering up the bones to be later burned or interred with the bones of other ancestors, or sometimes with the intent of stealing precious items left with the dead body—those dried remains would crumble into dust. In other cases when they returned before the body had dried out, particularly in more humid or wet climates, the body would often contain maggots. These routine observations therefore also connected death with dust and maggots.

Ancient Mesopotamian Thinking

Akkadian literature, the oldest that we have found to date, and predating that of the Hebrews by millennia, describes a world of the dead deep within the earth. Its entrance is found in the west (the direction of the sunset, which marks the ending of the day and of light), and entry is gained through seven successive sets of gates, the first of which is guarded by a watchman.[3] This "land without return" contained a palace for the divine ruler, and was a dark place filled with dust, as well as a stream of water flowing through it. Three narratives survive which vividly describe the afterlife as seen through their eyes. We have already described Gilgamesh's trip to the underworld in search of his beloved but now deceased friend, Enkidu, a story made more vivid by his observation of a maggot crawling out of Enkidu's nostril

---

3. Landman, "Sheol," 282.

(symbol of death in the conduit for the breath of life). He is briefly reunited with Enkidu, and this part of the story abounds with references to weeping, imprisonment, and dust. Another Mesopotamian narrative paints a haunting picture of the descent of the goddess Ishtar into the netherworld:

> To the house which none leave who have entered it,
> To the world from which there is no way back,
> To the house wherein the entrants are bereft of li[ght],
> Where dust is their fare and clay their food,
> (Where) they see no light, residing in darkness,
> (Where) they are clothed like birds, with wings for garments,
> (And where) over door and bolt is spread dust.[4]

A third narrative recounts Nergal (the plague god) usurping control over the underworld from the rule of Ereshkigal (queen of the Underworld): one version has him doing this through force, another version through marriage.[5]

A significant amount of the texts that the Egyptians left for us are referred to as "coffin texts": as their name implies, these covered the coffins of the more prominent members of Egyptian society, and provided instructions for the deceased to help guide them on their postmortem journey. We also have a voluminous Egyptian *Book of the Dead* (this featured prominently in the Indiana Jones movie *Raiders of the Lost Ark*). The Egyptians developed elaborate embalming techniques to maintain the integrity of the body, and laid the latter in richly decorated coffins placed in tombs furnished with foods and personal items (jewelry; weapons; statues).[6] The afterlife involved a whole society of gods, animals, and humans, the latter retaining the social rank they held in life. Death was the beginning of a new journey which was an extension of one's mortal life: the deceased would progress through different stages of the underworld, eventually reaching a great hall of judgment where their soul was weighed.[7] Those found to have passed judgment attained *akh*: an eternal existence among the stars; those that did not were discarded on the floor. (Might the author/editor of Job have had this imagery in mind when writing: "Let God weigh me in honest scales and he will know that I am blameless" [Job 31:6]?)

---

4. Madigan and Levenson, *Resurrection*, 44; Pinker, "Sheol," 172; Dalley, *Myths*, 154–64.

5. Dalley, *Myths*, 163–81.

6. Quirke, Exploring Religion, 201–37; Davies, Death, 92–95.

7. Johnston, *Shades of Sheol*, 69.

These two global superpowers defined the ancient Near Eastern understanding of the afterlife for thousands of years, at a time when Abram left on his journey to Canaan and the nation of Israel was forming (and creating its own traditions and writing its own religious texts). Those civilizations would undoubtedly have influenced the development of the latter, especially given that Moses (who may have written some of the earliest Hebrew texts, and some believe he wrote much of the Pentateuch) was a member of Pharaoh's household and was schooled in the ancient Egyptian academy, and that Hebrew scribes during the exilic period (who are credited with extensive redaction and revision of many Hebrew texts, including the earlier Mosaic ones) were immersed in the literature of both empires.[8] It will be noted later that Daniel's writings are the first (some claim his writings are *the only* ones) in the Hebrew Bible to explicitly introduce the concept of bodily resurrection for the faithful. The Zoroastrians of the Persian empire (where we find Daniel) had developed an elaborate concept of the afterlife in which one's spirit traveled to an underground kingdom (aided by mortuary blood sacrifices made by the surviving family) where it encountered a bodily representation of its good or bad deeds, then departed to some heavenly realm prior to being united with that bodily representation at a future date.[9] As such, the Zoroastrians may have contributed to the Hebrew development of a concept of bodily resurrection and a joyous form of afterlife in paradise.[10]

Ancient Israel's Death Cult?

Recent scholarly consensus posits that the religious beliefs of early Israel mimicked those of the surrounding nations, including their own death cult tradition, until an exclusively monotheistic Yahwism appeared in the late monarchic and exilic periods, and religious reforms were imposed during the reigns of Kings Hezekiah and Josiah.[11] The early Israelite death cult included: veneration of one's ancestors; invoking the names of the dead when making vows and at family feasts; consulting the dead and related forms of divination; and leaving food offerings to feed the dead. This claim is based upon archaeological evidence as well as veiled references and remnants of passages found in Old Testament books which appear to have escaped or

---

8. Landman, "Sheol," 282; Johnston, *Shades of Sheol*, 69.

9. Elledge, "Resurrection," 35; also see Martin-Achard, *Death to Life*, 190.

10. Stroumsa, "Paradise Chronotrope," 1, 3; Schaper, "Messiah in the Garden," 19.

11. Johnston, *Shades of Sheol*, 17; Bloch-Smith, *Judahite Burial*, 109–32; Madigan and Levenson, *Resurrection*, 55–62; Toorn, *Family Religion*, 206–35; Moreland and Rae, *Body and Soul*, 32.

survived the textual editing by postexilic scribes. For example, Bloch-Smith highlights passages which suggest that the ancient Hebrews attended to these needs of the dead out of a belief that the latter could: provide advice or wisdom to the living (King Saul consulting the witch of Endor to raise Samuel; 1 Sam 28:6–19); provide a fertility blessing (the annual sacrifice that Elkanah and his family made at Shiloh; 1 Sam 1); revive the dead (Elisha's bones reviving a corpse; 2 Kgs 13:20–21); and exact vengeance (King David cutting off the hands and feet of his already dead enemies; 2 Sam 4:12).[12] Several scholars have pointed out how Deuteronomic and Holiness Law Codes had to be invoked in the eighth and seventh centuries BCE in order to control a prevalent practice of feeding or consulting with the dead (Exod 22:18; Lev 19:31; 20:6, 27; Num 25:2; Deut 18:10–11; 26:14; Pss 16:3–4; 106:28) or of human sacrifice (Lev 20:2–5).[13] The repeated admonitions against these practices attest to the prevalence of the latter in earlier Hebrew history, and additional references to it in the prophetic books (Isa 8:19; 57:5–7, 9; Ezek 43:7, 9) suggest elements of this practice may have continued to some extent even into that later era.[14] Human sacrifice was prevalent in the nations surrounding Israel,[15] and there are instances recorded of it occurring in Israel as well (Judg 11:31, 39; 1 Kgs 16:34 [cf. Josh 6:26]; 2 Kgs 3:27). To be fair, though, other contemporary scholars argue equally *against* the idea that Israel at times practiced a death cult.[16]

Bloch-Smith documents a wide variety of items with which the dead were occasionally buried in ancient Palestine (although some dispute whether certain of the buried remains were Jewish or pagan), and conjectures that their relative lack of symbolic or emotional value "suggests that the living expected the dead to continue performing daily chores to meet their physical needs after death."[17] These items included jewelry (bracelets; amulets; beads; rings; earrings; pins; scarabs), spoons, pottery (bowls; vases; flasks; chalices; vessels for preparing, serving and storing food and wine), food, tools (lamps; blades; arrowheads; chisels; flints; hooks; fishhooks; grinding stones; weights; tongs; tweezers; pins; needles; loom weights; spindles), and personal items (comb; mirror; hair clasps; gaming pieces; stamps;

---

12. Also see Johnston, *Shades of Sheol*, 150–95; Toorn, *Family Religion*, 231–35.
13. Johnston, *Shades of Sheol*, 152–53, 169–70.
14. Bloch-Smith, *Judahite Burial*, 121; also see 129–32.
15. Johnston, *Shades of Sheol*, 35–37.
16. Madigan and Levenson, *Resurrection*, 62–66.
17. Bloch-Smith, *Judahite Burial*, 90.

seals).[18] Some were also buried with statues of fertility goddesses and other anthropomorphic beings of unknown relevance.[19]

Ancient Hebrew View of the Afterlife

The earliest mention in Hebrew Scriptures of the human experience upon meeting death is found in the opening chapters of Genesis, and is uttered by God himself: Adam was told simply that he would "return to the ground, since from it you were taken; for dust you are and to dust you will return" (Gen 3:19). There is no mention here by God of any conscious after-death existence of any kind—neither residence in some kind of place which is described as heavenly or hellish, nor any mention of a transformation into some new kind of body—and many other Old Testament passages remind us that death is followed by decay of the body without mentioning any other form of continued immaterial existence (Pss 90:3; 103:14b; Eccl 3:20). The author of Ecclesiastes does so in incredibly poetic fashion: "Remember [your Creator]—before the silver cord is severed, and the golden bowl is broken; before the pitcher is shattered at the spring, and the wheel broken at the well, and the dust returns to the ground it came from" (Eccl 12:7). Other passages reinforce the brevity of human existence, alluding to flowers or grass which spring up and exude life but then quickly dry up and disappear (Job 14:1; Ps 90:5; Isa 40:6). More importantly, many Old Testament passages affirm that, following this brief existence and then death, the soul then rests in Sheol, completely out of contact with the world of the living or with others who have also died. Job absolutely denies any thought of resurrection (Job 14:7–12). David also writes that the dead no longer enjoy any kind of fellowship with the Lord, and this view is echoed by King Hezekiah (Pss 6:5; 30:9; 88:11–12; 115:17; Isa 38:11,18,19; Sir 17:27–28). Some Psalmic passages suggest that their authors questioned whether there was any actual existence of any kind after death (Pss 6:5; 39:13; 41:5; 115:17). The writer of Ecclesiastes depicts an empty lonely existence for the departed (Eccl 3:18–19; 9:5–6).

These views of the afterlife are very similar to those of the Akkadians, Sumerians, Egyptians, and ancient Greeks. And yet the ancient Hebrews distinguished themselves from those surrounding nations by seeming to be not at all preoccupied with postmortem existence.[20] That is, the Hebrew root word for "die" or "death" (*m-w-t*) occurs one thousand times, but there are

18. Bloch-Smith, *Judahite Burial*, 63–94; also see Johnston, *Shades of Sheol*, 62–64.
19. Bloch-Smith, *Judahite Burial*, 94–103.
20. Madigan and Levenson, *Resurrection*, 66; Green, *Body, Soul*, 153.

only roughly one hundred references to the underworld, including only sixty-six occurrences of the proper noun Sheol,[21] and the most significant term for an inhabitant of Sheol (*rᵊpā'îm* or *rephaim*) occurs only eight times.[22]

Faint glimmers of an emerging belief in a more positive version of afterlife existence with YHWH can be found elsewhere in the Old Testament. Contradicting their earlier statements cited above, David also says: "God will redeem me from the realm of the dead; he will surely take me to himself" (Ps 49:15), while the writer of Ecclesiastes also says: "The dust returns to the ground it came from, and the spirit returns to God who gave it" (Eccl 12:7). Several other passages in Psalms hint at an eternal enjoyment of the presence of God (Pss 16:9–10; 73:23–26). Job also seems to contradict his own words by saying, "After my skin has been thus destroyed, yet in my flesh I shall see God" (Job 19:25), although he may be referring here to his full and complete physical recovery from the boils and painful sores which covered his entire body, rather than to resurrection from the dead.

Johnston claims that the distribution or localization of the word Sheol within the Hebrew texts—where it *is* used, and where it is *not* used—is particularly revealing. It is *not* used in narrative accounts of life histories (for example: "King X was a wicked king; he died and descended into Sheol"), and it is completely absent from their legal texts, including those which describe capital punishment. If there were to have been any references in the Old Testament to Sheol as punitive, it would be in such passages. Instead, it is found in passages involving intense personal engagement: in Psalmic literature twenty-one times; in reflective literature (Job, Ecclesiastes) twenty times; in prophetic literature seventeen times; and in narrative literature eight times.[23] In the latter case, it is used almost entirely in direct speech. For example, David, on his death bed and instructing Solomon how to tie up loose political strings, says about his enemy Joab: "Deal with him according to your wisdom, but do not let his gray head go down to Sheol in peace" (1 Kgs 2:6).

What, then, did the ancient Hebrews mean by this word that they used rarely and exclusively within such personally engaging passages?

Sheol: A Very Gloomy Place

The word "Sheol" is never accompanied by the definite article ("the"), suggesting it is a proper name.[24] The etymology of the word is uncertain.

---

21. Johnston, *Shades of Sheol*, 72.
22. Johnston, *Shades of Sheol*, 127–28; Moreland and Rae, *Body and Soul*, 31.
23. Johnston, *Shades of Sheol*, 71–72.
24. Johnston, *Shades of Sheol*, 71.

It had previously been thought to be an Assyro-Babylonian loan word *Shu'alu*—having the assumed meaning of "the place where the dead are cited or bidden," or "the place where the dead are ingathered"—but that idea has since been abandoned.[25] An Aramaic papyrus from Elephantine uses the word *sehol*, albeit only once, in the context of the grave and a final location for one's spirit ("thy bones shall not go down to *sehol*, nor thy spirit"), but Bar cautions against attributing too much to this singular occurrence of use.[26] Johnston critiques a number of suggested links to Semitic and non-Semitic languages, dismissing all but one which finds a reference to an underworld deity in an Akkadian text, likely Ereshkigal.[27]

Hebrew synonyms for Sheol include *bor* (pit), *abaddon* and *shahat* (pit or destruction) and possibly also *tehom* (abyss).[28] Johnston also goes through a very lengthy lexical analysis of whether the word *eretz*—which appears frequently and is usually translated as earth, world, land, country, or ground—might also mean underworld. He then repeats that exercise with an equally lengthy analysis of whether the occurrences of "water" can also refer to the underworld.[29] His conclusion is that "water" and "earth" may be associated with the underworld, but are not equated with it.

Paradoxically, Sheol is described in Scripture as a land of no return (2 Sam 12:23; 14:14; Job 7:9, 10; 10:21; 14:10–12; 16:22; Jonah 2:6; Ezek 26:20), and yet everything we know about it comes from the testimonies of the righteous who claim to have in fact been brought back up and redeemed from Sheol (1 Sam 2:6; Pss 18:16; 30:3; 40:2; 49:15; 71:20; 86:13; 107:10–22; Jonah 2:2, 6), the most vivid account of which is found in Psalm 88.[30] Another paradoxical claim about Sheol is that it is a place where YHWH is no longer remembered (Pss 6:5–6; 115:17; Isa 38:18), and yet many other passages describe those trapped in Sheol as crying out to him (Pss 31:17; 88:9, 13; 107:19–20; Jonah 2:1–9).

Several Old Testament texts refer to Sheol as a destination or place, the journey to which is always downward (Num 16:30, 33; 1 Sam 2:6; Job 17:13–16; Pss 49:17; 55:15; Prov 5:5; Jonah 2:3–8; Isa 14:11, 15; 38:18; 57:9;

---

25. Landman, "Sheol," 282.

26. Bar, "Grave Matters," 147; also see Harris, "Meaning of the Word Sheol," 129; Johnston, *Shades of Sheol*, 77.

27. Johnston, *Shades of Sheol*, 78–79.

28. Martin-Achard, *From Death to Life*, 37; Landman, "Sheol," 282; Bar, "Grave Matters," 145; Harris, "Meaning of the Word Sheol"; Pinker, "Sheol," 171; Johnston, *Shades of Sheol*, 83–85.

29. Johnston, *Shades of Sheol*, 101–23.

30. Madigan and Levenson, *Resurrection*, 48, 66; also see Crenshaw, "Love Is Stronger," 54–55.

Ezek 31:14–17; 32:21–32) and ending in some place very deep (Ps 88:6; Prov 15:24), far deeper than just the shallowness of a grave. In fact, Sheol is often given as the physical antipode of heaven itself (Job 11:8; Ps 139:8; Isa 7:11; Amos 9:2). On the other hand, escape from Sheol is always upward (1 Sam 2:6; Pss 30:3; 86:13; 88:10; Jonah 2:6).

Descriptions of Sheol often evoke a dry and dusty underground existence—often being translated as the "pit" (Pss 30:3; 88:4, 6; Isa 14:15; Ezek 31:16; 32:23, 25, 30) or the "grave" (Job 24:19; Pss 88:5, 11; 89:48; Prov 5:5; 7:27; Isa 14:11; Ezek 32:23)—but sometimes an abysmal watery existence (2 Sam 22:5; Ps 88:7; Jonah 2:3–8). Surprisingly, these diametrically opposite existences can be found in the same passage (for example, Ps 88). Crenshaw traces these two quite different characterizations back to the Paleolithic and Neolithic periods, claiming that wandering semi-nomads believed their dead rested in graves as "living corpses," while city dwellers in ancient Mesopotamia envisioned their dead existing in a watery underworld.[31]

Sheol is an unpleasant place, a "land of oblivion" (Ps 88:12) and destruction (Ps 88:11), one associated with: darkness (Job 10:21; Ps 88:6, 12; Lam 3:6); silence (Pss 31:17; 115:17; Isa 38:18); decay (Ps 16:10); worms and maggots (Job 17:13–16; 24:19–20; Isa 14:11); and dust (Job 17:16). It is compared to a prison with seven gates (Job 17:16; 38:17; Pss 9:14–15; 88:8; 107:16–18; Isa 38:10) and suggested to have compartments or cells (Prov 7:27); elsewhere, its metaphor is a snare with coils (2 Sam 22:6; Ps 18:4), or some kind of monster that opens up its jaws and swallows people whole (Exod 15:12; Num 16:32–34; Isa 5:14; Hab 2:5).

Sheol: Ultimate Destiny for All?

Sheol is almost always associated with the dead and with death (Job 17:13–16; Pss 30:3; 49:14, 15; 86:13; 88:5; 89:48; Prov 5:5; 7:27; Isa 14:15; 38:10; Ezek 31:15, 16; 32:21). Yet certain of the wicked can go down to it alive (Num 16:23–33; Ps 55:15), and others in Sheol continue some form of existence, although they are mere shadows (*rephaim*) of what they once were (Isa 14:9; 26:14; Ps 88:4), weak (Isa 14:9–10; 59:10), inactive (Eccl 9:10; Ezek 32:22–30) and otherwise sleeping (1 Sam 28:15; Job 3:13; 14:12; Ps 13:3; Jer 51:39, 57).

The psalmist asks rhetorically whether Sheol is not the ultimate destiny for *all* humans (Ps 89:48–49), and Qoheleth answers in the affirmative (Eccl 9:7–10). This had in fact been the traditional understanding of modern scholars up until the past few decades. In the words of one prominent

---

31. Crenshaw, "Love Is Stronger," 54.

scholar: "The ideas of grave and Sheol cannot be separated. . . . Everyone who dies goes to Sheol."[32]

Other Old Testament passages describe Sheol as a place for the wicked (Pss 9:17; 31:17b; 55:15; Job 24:19). However, the correspondingly alternative place for the righteous is never mentioned in the Old Testament, even though the first few chapters in Genesis have already introduced an ideal counterpart in the form of the garden of Eden or paradise. Moreover, two of the righteous were seemingly spared death entirely, but their final destiny is not detailed: Enoch is simply "taken away" and "was no more" (Gen 5:24), while Elijah "went up to heaven in a whirlwind" (2 Kgs 2: 11). Nor do any other Old Testament authors or characters ever indicate their wish to end their days on earth in the same manner as was described for either Enoch or Elijah. Surprisingly, even some of the righteous do sometimes find themselves in Sheol, or at least dread being found there (Gen 37:35; Job 17:13-16; Pss 31:17; 88:4; Isa 38:10).

This concept that Sheol was the common destiny for all—both the wicked and the righteous—undoubtedly fomented frustration and tension in the minds of the ancient Hebrews: a sense of injustice that even the most righteous would share the same destiny as the most wicked, and that YHWH would demand justice from people but not practice it himself; the indignity that the devout practitioner of the covenant would lie down forever with the uncircumcised (Ezek 32:19-32). This certainly contributed to the development of the idea of a postmortem judgment and eternal recompense for one's life on earth. This development will be explored below.

The "Bad Death" of the Wicked

There are far more biblical references to Sheol being a destination for the wicked than for the righteous.[33] In one of the most vivid and horrifying Sheol accounts—that of Korah and his followers receiving Divine judgment for their act of rebellion—the earth opens up beneath them and swallows them and their families alive into Sheol. Many prophetic passages speak of Israel's enemies being cast down to Sheol (Isa 14:11, 15; Ezek 31:15-17; 32:18-32).

---

32. Pedersen, *Israel*, 461; also see Martin-Achard, *From Death to Life*, 38-39; Madigan and Levenson, *Resurrection*, 42-43; Bar, "Grave Matters," 146; Harris, "Meaning of the Word Sheol."

33. Johnston, *Shades of Sheol*, 81.

Bar claims, "Sheol is not simply the place where wicked people go, but rather the place of a 'bad death.'"[34] Johnston defines what a "bad death" is: "Normally death is natural, but it is bad when premature, or violent, or there is no surviving heir."[35] Old Testament passages which point to the disgrace of a premature or particularly violent death include 2 Sam 3:33 and Eccl 7:17. A "bad death" could also pertain to the body being desecrated and/or left unburied and accessible for birds and animals to mutilate, as might occur on a battlefield (Deut 28:25; 2 Sam 4:12; Jer 7:33; 8:1; 16:4, 6).

A hallmark feature of the descent into Sheol is a feeling of being "cut off" and forgotten. Those in Sheol feel abandoned by YHWH, and are unable to praise or "remember" YHWH (Pss 6:5; 88:5, 10–14; 115:17; Isa 38:18). Adding to that, they feel separated from their loved ones and the community (Ps 88:5, 8; Isa 26:14).[36] The account of Korah and his families being thrown alive into Sheol concludes with the simple statement "they perished and were gone from the community" (Num 16:30–34). Likewise, one of the most heart-rending accounts of the feeling of separation in Sheol is given by Job after losing all of his children and feeling like he had also lost favor with YHWH, a soliloquy in which he paints this image of permanent hopelessness and being forgotten:

> Consider that my life is but wind; I shall never see happiness again.
>
> The eye that gazes on me will not see me; your eye will seek me, but I shall be gone.
>
> As a cloud fades away, so whoever goes down to Sheol does not come up;
>
> He returns no more to his home; his place does not know him. (Job 7:7–10).

The sense of separation and the descent into Sheol do not have to be sudden transitions, nor total and abrupt discontinuities: several Old Testament passages describe these as partial, progressive, pernicious and malignant, much like a grave illness which can bring on that alienating experience. This certainly seems to be in view in Psalm 88. When one becomes deathly ill, their world is reduced to the confines of the bedroom, to just the bed, even just the pillow on which one's head lies, and everything else from the "outside world" recedes from relevance as energy and life slowly

---

34. Bar, "Grave Matters," 148; also see 151.
35. Johnson, *Shades of Sheol*, 41.
36. Johnston, *Shades of Sheol*, 32.

ebb away, until one finally dies: this too can be the experience of descending into Sheol. In the same way, certain Old Testament passages describe the plaintiff as hovering at the edge of the Pit (Ps 88:3–4), or just beginning to experience the hopelessness and abandonment of Sheol.

Mitchell weaves a very interesting (and convincing) tapestry around "the Korahite Psalms" and their "affiliates" (Pss 42–49 and 84–89). Although there are several Korahs in the Bible, he argues that these psalms were written by the sons of the only one so named whose story is told in detail: a Levite (1 Chr 6:22) and cousin of Moses and Aaron who resented Moses' decision to appoint headship of the Kohathite clan to a lesser relative, and who finally led a rebellion against Moses' leadership during the exodus from Egypt (Num 16).[37] Korah and 250 followers—together with their families and possessions—were swallowed up by a sudden opening in the earth and consumed by a fire which "came out from the Lord." Later, we read of Korah's sons, who had presumably separated themselves from their father's evil leadership and joined with the rest of the Israelites when divine judgment was being meted out (Num 26:9–11, 58). Those sons of Korah are presumed, then, to be the authors of the Korahite Psalms which have redemption from Sheol as a major theme: they contemplated how they too would otherwise have been destined for destruction in Sheol with their father, except for the redemptive act of YHWH in response to their wise decision (Prov 15:24) to separate themselves from evil. Mitchell adds some color to the stories of two other descendants of Korah: an interesting observation regarding Hannah, the mother of the prophet Samuel who in turn was the only person in Hebrew Scripture to have been briefly returned from Sheol (1 Sam 28:4–19),[38] as well as an explanation for the very puzzling story of the forty-two boys who were mauled by a bear for taunting the prophet Elisha for being bald (2 Kgs 2:23–24).[39]

Sheol, then, can be a status or condition of perceived and seemingly permanent separation from YHWH, ancestors, descendants, and community; the unfortunate can find themselves cut off (due to sickness, war, punishment, judgment, or some other calamity) and call out to YHWH for rescue and restitution; the wise can make personal choices which spare themselves from being cut off in Sheol (Prov 15:24), or they can discipline their children in order to spare them from the same fate (Prov 23:14); the unwise or wicked, on the other hand, can choose to have nothing to do

---

37. Mitchell, "God Will Redeem," 368.
38. Mitchell, "God Will Redeem," 371.
39. Mitchell, "God will redeem," 373–74.

with YHWH, or to do something that alienates them from ancestry, descendants or community.

The "Good Death" of the Righteous

In contrast to the "bad death" of the wicked, what does a "good death" look like? Against the many passages which portray a very negative response to the isolation and abandonment to be found in Sheol—an experience that the living would seek to avoid (Job 17:13–15; Prov 15:24; Isa 38:10) or wish upon one's enemies (Pss 9:15–17; 31:17; Ezek 31:15–17; 32:21–27)—certain other passages depict the righteous calmly accepting death, even inviting it, and being reunited with their ancestors. Jacob delivers a long eulogy in which he blesses each of his sons, then resolutely "drew his feet up into the bed, breathed his last and was gathered to his fathers" (Gen 49:33). This account echoes the ones pertaining to the deaths of his father Isaac and grandfather Abraham, both of whom seemed to have calmly "breathed his last" before dying "old and full of years" and being "gathered to his fathers" (Gen 25:7–8; 35:28–29). Many others were said to have been "gathered to his fathers" or "slept with his fathers" in a way that denoted a "good death," including Ishmael (Gen 25:17), Moses (Num 27:13; 31:2; Deut 32:50), Aaron (Num 20:24; 27:13; Deut 32:50), Gideon (Judg 8:32), David (1 Kgs 2:10; 1 Chr 29:28), Solomon (1 Kgs 11:43), Jehoram (2 Kgs 8:24), Josiah (2 Chr 34:28), Jehoiakim (2 Kgs 24:6), and Jotham (2 Chr 27:9).[40] In many cases, the setting for this wording includes looking back on a full and long life, with several generations in view honoring and remembering the departing individual, and a relatively peaceful death bridging the divide between the living descendants and the dead ancestors: there is very much the idea of family and multi-generational continuity. In fact, starkly contrasting his peaceful acceptance of death referred to at the beginning of this paragraph when he was surrounded by his sons, Jacob had just a few years prior exhibited a very negative response toward the prospect of death when he thought that Joseph had been killed by a wild animal (Gen 37:34–35) and later when he feared he might also lose Benjamin (Gen 42:38; 44:29, 31).

As stated above, then, the ancient Hebrews seemed not to be preoccupied with death: for them, it was simply a part of the experience which connected them with their ancestors. Suriano provides an excellent account of how this perspective is reflected in the ritual of ancient Hebrew burial

---

40. It is worth emphasizing that these references always point to fathers and sons, not the more inclusive "people."

and in the architecture of Judahite bench-tombs.[41] The latter were prevalent in the hill country where large, open, and flat parcels of land were not as available as in the lowlands (where bodies were generally buried in the ground). The bench-tombs were familial rooms or caverns hewn into rock which served to control the chaotic forces of death. Upon someone's death, the family would wash the body then carry it out to the family tomb and lay it on the bench inside, close to where the family's ancestral bones were kept (the individual "slept with his fathers"). After placing various grave goods with the body, the tomb was sealed for a period of time while the chaotic forces devoured the flesh leaving only the dry bones. Later, the living relatives would go back in and transfer those dry bones to a jar or ossuary which held the dry remains of previous generations (the individual was "gathered to his fathers"). Each of these steps was accompanied by extensive liturgy. This ritual connected the generations of the living who carried the body into the tomb with the generations of the dead who resided there and received it. It is worth noting that the Israelite burials in the Late Iron Age always left the body fully extended and on its back (as if in sleep) and never curled in a fetal position (as if in preparation for a rebirth) or on its side (as if to face a new day);[42] this contrasts with the surrounding nations, and with early Israel (see section on Ancient Israel's Death Cult above).

This desire for ancestral connection is likely behind the careful preparations made by Abraham and Joseph for their own final resting places. Abraham was driven to secure a tomb for his wife Sarah, one that would be completely separate from those of the Hittite people with whom he was living at that time (Gen 23). This was a deliberate and protracted negotiation with the Hittites, not an afterthought: considerably more textual space is given to the negotiations around the purchase of the land than the actual burial of Sarah itself. This same tomb later received the bodies of Abraham himself, his son Isaac, Isaac's wife Rebekah, his grandson Jacob, and Jacob's wife Leah (Gen 25:9–10; 49:29–30; 49:31–32; 50:12–14); it is interesting that Rachel, the wife whom Jacob loved most, was not buried there (Gen 35:19–20). Joseph, who held a very high-profile position in Egyptian society (which had its own very specific beliefs and rituals surrounding burial), is said to have explicitly instructed his descendants to carry his bones from Egypt when God liberated them, even making them swear to this (Gen 50:24–26; Exod 13:19). Later they are duly recorded as having followed through on that oath (Josh 24:32; note, cf. Gen 33:18–20 which describes

---

41. Suriano, "Sheol, the Tomb," 29; also see Bloch-Smith, *Judahite Burial*, 41–52, 136–37; Johnston, *Shades of Sheol*, 57–64.

42. Johnston, *Shades of Sheol*, 69.

an earlier purchase of land which is relevant to this story). Centuries later, Stephen is speaking to a Hellenistic audience of Jews and recounts a factually different story (Acts 7:15–16): his speech raises questions about who exactly bought the land, how many purchases were involved, and what is the location of the purchased lands. Several explanations have been given for this discrepancy, including: the common practice employed of telescoping history and bending context; the political tensions around Shechem being a major Samaritan settlement (the religious enemies of Stephen's audience); and even Stephen as speaker or Luke as historian making a historical mistake. Those hermeneutical problems and solutions are beyond the scope of this essay. Nonetheless, it is relevant here (and not disputed) that Abraham and his descendants were very intent upon having their bones being kept in a familial location.

Resurrection and the Afterlife in Second Temple Judaism

Following such a lengthy discussion of death and Sheol, it would seem natural to proceed next to a discussion of what some might think would be Sheol's opposite(s): the garden of Eden and paradise. However, those two "places" presuppose the concept of a postmortem life. So we must first explore resurrection before moving on to an afterlife existence *with* God; furthermore, before proceeding to the latter topic, we will explore how the Hebrew concept of Sheol transformed into the Greek one of Hades.

The two different views regarding Sheol—it being on the one hand a universal and permanent destiny for all humans, and on the other hand a reversible or avoidable separation from loved ones—are not mutually exclusive. However the first of these two is more likely to necessitate the later development of the concept of resurrection, final judgment, and distinct destinies for the righteous and the wicked. That is, the sense of injustice and indignity of a universal or common Sheol would force difficult questions for a devout Hebrew:[43] they might ask "How could YHWH allow the righteous Jew to share a common eternal destiny with the uncircumcised and the wicked? What is the point of serving YHWH while alive on earth if he is going to discard and disregard us forever?" These kinds of questions resonated with other ones such as that with which Job wrestled: can the righteous expect a problem-free life and the "wicked" to suffer?

Several prominent biblical figures exhibit double-mindedness on this theological question. Job absolutely denies any thought of resurrection

---

43. Bar, "Grave Matters," 146; Pinker, "Sheol," 178.

(Job 14:7–12), but later seems to contradict himself by saying, "After my skin has been thus destroyed, yet in my flesh I shall see God" (Job 19:26) (although, as already noted above [page 136], he may be referring to a more common recovery from his skin affliction rather than to a very uncommon resurrection from the dead). The psalmist writes that the dead no longer enjoy any kind of fellowship with the Lord (Pss 6:5; 30:9; 88:11–12; 115:17), but also writes: "God will redeem me from the realm of the dead; he will surely take me to himself" (Ps 49:15), and elsewhere hints at an eternal enjoyment of the presence of God (Ps 16:9–10; 73:23–26).[44] The writer of Ecclesiastes writes:

> As for humans, God tests them so that they may see that they are like the animals. Surely the fate of human beings is like that of the animals; the same fate awaits them both: as one dies, so dies the other. All have the same breath; humans have no advantage over animals. Everything is meaningless. All go to the same place; all come from dust, and to dust all return. Who knows if the human spirit rises upward and if the spirit of the animal goes down into the earth? (Eccl 3:18–19)

But shortly thereafter, that author also writes: "The dust returns to the ground it came from, and the spirit returns to God who gave it" (Eccl 12:7) and later hints at a final judgment (Eccl 12:14). So the biblical texts preceding the writing of the book of Daniel are not at all clear on the subject of resurrection.

The cognitive dissonance produced by this theological matter—the faithful Jew sharing the same final destiny as the Gentile—would unsettle Hebrew minds for centuries, and would reach a fever pitch during the exile when they were being persecuted and even martyred for their belief and faith practices.[45] This set the stage not only for a paradigm change in their understanding of the afterlife, but also introduced the concepts of resurrection, postmortem judgment, heaven, and hell. These concepts all developed considerably during the Hellenic and Hellenistic periods.[46]

Charlesworth provides fifteen different categories of "resurrection" in Scriptures.[47] Here we will consider only resurrection to an entirely new

---

44. Although these may be later editorial insertions added after Daniel has introduced the concept of resurrection.

45. Crenshaw, "Love Is Stronger," 56–57.

46. Crenshaw, "Love Is Stronger," 71–72; Madigan and Levenson, *Resurrection*, 5; D'Costa, *Resurrection Reconsidered*, 188–89; Davies, *Death, Ritual and Belief*, 119; Scott and Phinney, "Relating Body and Soul," 92; Scharen, "Gehenna," 324–37.

47. Charlesworth, "*Where Does the Concept*," 2–17.

existence. This excludes many instances of dead bodies being resuscitated and returned to a normal earthly existence in which the body continued to grow old and ultimately died once again (without another recovery). The latter group include: the son of the Zarephath widow (1 Kgs 17:17–22), the son of the Shunammite woman (2 Kgs 4:32–35), the man whose dead body touched Elisha's bones (2 Kgs 13:20, 21), the son of the widow of Nain (Luke 7:11–15), Lazarus (John 11:1–44), Jairus's daughter (Luke 8:41, 42, 49–55), many dead saints in Jerusalem when Jesus was resurrected (Matt 27:50–53), Tabitha, whose Greek name was Dorcas (Acts 9:36–41), and Eutychus (Acts 20:9, 10). It is also important to distinguish between resurrection of the physical body versus being given a new, very different kind of body (for example, an immaterial or "spiritual" one).

Many scholars agree that the first clear reference in Hebrew Scripture to a physical, bodily resurrection to an entirely new existence is found in Daniel 12, written in the third or early second century BC.[48] In contrast to the sleep-like state or gloomy nonexistence in Sheol described in preexilic Old Testament texts, Daniel introduces a glorious existence for a select few who prove themselves faithful in persecution and martyrdom (Dan 12:1–3, 13): these saints will shine like stars in the heavens for eternity. In fact, Collins calls it "the *only* generally accepted reference to resurrection in the Hebrew Bible,"[49] while Charlesworth calls it "the only undisputed passage in the Old Testament of [this],"[50] and Johnston exclaims "here *(at last)* in the Old Testament there is a developed concept of resurrection" (italics added).[51] Many scholars attribute this sudden crystallization of Hebrew thinking on the subject of resurrection to Persian Zoroastrianism, which Israel encountered during the exile (see pages 11–12, and 133).[52] Davies argues that "Zoroastrians developed a belief in a more joyous form of afterlife through the resurrection of the body; in fact, it is likely that this was the first religious tradition to arrive at this route to a realm of Paradise."[53]

The vision of "the End Times" in the book of Daniel foretells a turbulent period of intense conflict between world superpowers vying for supremacy, and including brutal persecution of the Hebrew faithful,

---

48. Madigan and Levenson, *Resurrection*, 171–200; Johnston, *Shades of Sheol*, 225–27; Elledge, "Resurrection of the Dead," 25; Crenshaw, "Love Is Stronger," 65–66.

49. Collins, *Commentary*, 392; also see 394.

50. Charlesworth, "Where Does the Concept," 12.

51. Johnston, *Shades of Sheol*, 226.

52. Martin-Achard, *Death to Life*, 187–95; Elledge, "Resurrection of the Dead," 25, 35; Charlesworth, "Origin and Development," 221–22; Collins, *Commentary*, 394–96; Madigan and Levenson, *Resurrection*, 147, 175, 199.

53. Davies, *Death*, 82.

a time of distress such as has not happened from the beginning of nations until then. But at that time your people—everyone whose name is found written in the book—will be delivered. Multitudes who sleep in the dust of the earth will awake: some to everlasting life, others to shame and everlasting contempt. Those who are wise will shine like the brightness of the heavens, and those who lead many to righteousness, like the stars for ever and ever. (Dan 12:1–3)

The book of Daniel closes with a resurrection promise to Daniel also: "As for you, go your way till the end. You will rest, and at the end of the days you will rise to receive your allotted inheritance" (Dan 12:13).

Isaiah, also writing from within the Babylonian exile, hints at the concept of resurrection: "But your dead will live, Lord; their bodies will rise—let those who dwell in the dust wake up and shout for joy" (Isa 26:19). However, the context of this passage suggests it refers to the rebirth and renewal of the nation of Israel rather than resurrection of individuals to an eternal existence, and there is dispute as to whether it should be read literally or metaphorically.[54] The exact same points can be made for Ezekiel's vision of the valley of dry bones (Ezek 37).

This is an entirely new view of the afterlife than the one held by Hebrews for thousands of years in the ancient Near Eastern context. Certain of the righteous are now seen as being relocated to a better place and rewarded for their faith and actions done while in the body. Rather than resting in Sheol, cut off from the world of the living and even from YHWH himself, some could now look forward to shining like stars in the heavens for eternity. Stars were associated with angels, and thus this seemed to be a promise of an immortal heavenly existence.[55] The fate of the wicked, however, is simply "shame and everlasting contempt": there is no mention of flames or torment. Johnston points out that "the Hebrew Bible never indicates any form of punishment after death."[56] It is worth pointing out once again, that these two places of eternal postmortem existence are not automatically equated with the garden of Eden, paradise, or Sheol.

Resurrection in the book of Daniel is not universal:[57] the author(s) refer to "many who sleep," not "all who sleep," and in particular identifies "those who instruct in wisdom" and "those who turn the multitudes to righteousness" as

---

54. Collins, *Commentary*, 392, 395; Johnston, *Shades of Sheol*, 222–24.
55. Collins, *Commentary*, 393.
56. Johnston, *Shades of Sheol*, 73.
57. Collins, *Commentary*, 393; Johnston, *Shades of Sheol*, 226; Elledge, "Resurrection of the Dead," 27.

being the ones to be resurrected to everlasting life. According to Johnston: "The resurrection envisaged is not a general one of all humans, but focuses on the Jewish people, and possibly only one generation among them."[58]

This, then, is the beginning of a new emerging view of the afterlife. It is noteworthy that the eight occurrences of the term for an inhabitant of Sheol (r'pā'îm or *rephaim*) are found in only four Old Testament books (Job, Psalms, Proverbs, Isaiah), all of which are generally dated as exilic or postexilic.[59] This new view is then markedly developed during the Hellenistic period within apocryphal literature, especially in the *Books of Maccabees* 1, 2 and 4, and in *1 Enoch* 2–27; 92–105).[60]

The *Books of Maccabees* 1, 2, and 4, also written from within the Babylonian exilic context, present a slightly different view of individual resurrection. They contain testimonies of many believers not only *accepting* horrible torture and death, but even *seeming to pursue* the latter, because of their firm conviction that they will be rewarded in the afterlife for their suffering (2 Macc 14:46). These books promote a restoration of the physical body, with severed appendages reattached, rather than Daniel's view of resurrection to an astral existence with a new kind of body.

Many other Hebrew writers of the second and first centuries BCE were inspired by the books of Daniel and Maccabees to explore the afterlife and resurrection (1 Enoch 22–27; 92–105). Elledge focuses particularly upon two of the Dead Sea scrolls—*Pseudo-Ezekiel*, and *On Resurrection*, also known as the *Messianic Apocalypse*—both dating to the early first century BCE as setting the stage for Christian thinking on this subject.[61] *Pseudo-Ezekiel* reinterprets Ezekiel 37—the metaphorical resuscitation of dry bones to produce a new nation of Israel—as referring to the resuscitation of dead individuals. This physical interpretation of resurrection is quite unlike that promoted by Daniel (but is more like that in *2 Maccabees*). As for *On Resurrection*, Elledge claims it "is the only manuscript we have, dating prior to the origin of Christianity, that mentions both Messiah and resurrection within the same immediate context."[62]

Is it a coincidence that this novel Hebrew view on what happens to the believer after death appears at this time in Jewish history? Whereas the older view of resting in Sheol reflected that of the ancient Near Eastern culture

58. Johnston, *Shades of Sheol*, 226.

59. Johnston, *Shades of Sheol*, 128.

60. Charlesworth, "Where Does the Concept," 13; Crenshaw, "Love Is Stronger," 65–67; Davies, *Death*, 119; Johnston, *Shades of Sheol*, 229–30.

61. Elledge, "Resurrection of the Dead," 32–33; also see Charlesworth, "Origin and Development," 224; Charlesworth, "Where Does the Concept," 14–15.

62. Elledge, "Resurrection of the Dead," 32–33.

in which they were immersed, this newer view suddenly appeared within an entirely different zeitgeist, one which now included a serious threat to personal security: the Babylonian exile.[63] All of the books referred to in the preceding paragraph promise a reversal of fortunes (suffering turned to glory; restoration to newness of life) intended to console and encourage the faithful to remain true in the face of persecution, even to the point of brutal martyrdom. According to one scholar: "The catalyst that broke these ideas open and produced full-blown concepts of immortality and resurrection was apocalyptic theology, and its accompanying persecution of the righteous. The driving force and intellectual dynamic was the problem of theodicy."[64] Another explanation is that this novel view was also adapted from the surrounding culture. According to Collins: "It is hardly coincidental, however, that notions of astral immortality were current in the Hellenistic world."[65] "Because several of Israel's neighbors had well-developed doctrines of the afterlife for many centuries before the Jews, it is reasonable to suppose that foreign influence played a part."[66] In particular, Collins and others explore possible influences from Zoroastrianism as well as Greek, Babylonian, and Persian writings.[67] Alternatively, one could hold that this novel theological view was revealed by God at a unique time when Daniel and other believers particularly needed it.

Jesus and Paul Redefine Resurrection

Hendrikus Boers opens up his chapter on Christ's resurrection with an interesting point: "Paul was as little a Christian as Jesus had been. Both contributed fundamentally to the religion that later became known as Christianity, but neither of them were Christians: both were Jews. . . . When we make use of Christian language in interpreting him, we unavoidably read back into him ideas that had not yet developed in his thinking."[68] Some readers may find this statement to be very puzzling: how can Christ himself not be Christian? The point being made here is that Christianity became something

---

63. Collins, *Commentary*, 396.

64. Crenshaw, "Love Is Stronger," 71–72; see also Madigan and Levenson, *Resurrection*, 5; see also D'Costa, *Resurrection Reconsidered*, 188–89; see also Davies, *Death*, 119.

65. Collins, *Commentary*, 394.

66. Collins, *Commentary*, 396.

67. Madigan and Levenson, *Resurrection*, 175; Martin-Achard, *Death to Life*, 188; Elledge, "Resurrection of the Dead," 35; Charlesworth, "Origin and Development," 221–22.

68. Boers, "Meaning of Christ's Resurrection," 104.

very different from the message that Jesus preached, tremendously so in the first four or five centuries after Christ, and even more so leading up to the present time. We need to be very careful that we do not impose contemporary Christian ideas and interpretations onto *Christ's* teachings on resurrection and the afterlife.

By the time of Christ, we see two very different Jewish schools of thought on this question of the afterlife: the Sadducees who do not believe in it, and the Pharisees and Essenes who do (Acts 23:8).[69] Among these various views of resurrection which arose during the Hellenistic period, Jesus seemed to most accept the one introduced by Daniel. In many passages he predicts that he will rise again three days after being killed (Matt 12:40; 16:21; 17:22; 26:61; 27:63; Mark 8:31; 14:58; Luke 9:22; John 2:19). Immediately before those events, he tells his disciples that he is leaving to prepare a place for them (John 14:1-3), and shortly thereafter he tells another believer facing death with him: "Today you will be with me in Paradise" (Luke 23:43). After his death, he told Mary Magdalene to "not hold on to me, for I have not yet returned to the Father" (John 20:17). All of these are consistent with Daniel's view of resurrection to a new existence in a heavenly place.

The gospels and the book of Acts were written *after* Paul wrote his own letters and books, and for that reason one should keep open the possibility that their authors were influenced by Paul's teaching on this matter. Three of the gospels emphasize the *physical* nature of Jesus's resurrection. Matthew recounts that "Mary Magdalene and the other Mary . . . clasped his feet and worshipped him" upon seeing him at the empty tomb (Matt 28:8-9). Luke describes Jesus inviting the Eleven to touch his hands and feet to see that he is real, and not a ghost (Luke 24:37-40); later he eats a piece of broiled fish to prove that he has a physical body. John, on the other hand, describes Jesus commanding Mary Magdalene to "not hold on to me, for I have not yet returned to the Father" (John 20:17; unlike Matthew's version of this story, there is no mention of her actually touching him, nor of "the other Mary"); later, however, he appeared to the disciples and "showed them his hands and his side" (John 20:20) to prove his existence, and then invites Thomas to put his finger in the nail wounds in his hands, and to put his hand into Jesus's side into which the spear was thrust (John 20:27).

That having been said, this was not merely a resuscitation of Jesus's dead body to resume a life on earth: instead, this was something quite different. Shortly after his resurrection, eyewitness accounts describe Jesus being taken up into heaven and taking a seat at the right hand of God (Mark 16:19;

---

69. Crenshaw, "Love Is Stronger," 68-69; Johnston, *Shades of Sheol*, 230; Elledge, "Resurrection from the Dead," 37-41; Collins, *Commentary*, 398; Madigan and Levenson, *Resurrection*, 3; Davies, *Death*, 119.

Luke 24:51; Acts 1:9, 10).[70] Other passages in the gospels hint at Jesus having a non-physical body. After making a seemingly physical appearance to Mary Magdalene, he appeared to two other disciples "in a different form" (Mark 16:12; note, this passage and its surrounding verses are not found in the most reliable early manuscripts, which leaves open the possibility that they were added much later by some other editor/author). These two other disciples may have been the ones who Jesus encountered on the road to Emmaus, who Luke describes as somehow failing to recognize Jesus as he walked and talked with them throughout a day-long journey (Luke 24:13–16). Later that evening, they did recognize him as he broke bread, at which point he disappeared from their sight. Then he suddenly appeared standing among the disciples gathered in a closed locked room (Luke 24:36; John 20:20), inviting speculation that he could pass through doors or walls. All of these are things that a normal physical body does not do (although see the footnote on page 176 which points out that John does *not* say that Jesus walked through the walls of this locked room).

It is relevant to note here the account of the transfiguration of Jesus in the three Synoptic Gospels (John does not have this anecdote). Mark, the earliest of the gospel accounts, says nothing about Jesus's body itself, but comments only that "his clothes became dazzling white, whiter than anyone in the world could bleach them" (Mark 9:3). The accounts in the books of Matthew and Luke, on the other hand—both of which many scholars believe were written long after the book of Mark, and may have even borrowed from Mark's gospel account—expand this anecdote to also say that not just his clothing but also "his face shone like the sun" (Matt 17:1–13; Luke 9:28–36). Although these three accounts are not referring to resurrection specifically, the fact that Jesus is described as having dazzling brightness and talking to Moses (who died millennia before on Mount Nebo) and to Elijah (who is said to have been "taken" directly into heaven rather than dying) does recall Daniel's vision of certain resurrected believers shining like stars in the heavens. Likewise, the dazzling brightness of the angel(s)

---

70. I do not see heaven as being a physical place, let alone a place hovering above the Middle East, so this imagery of him being taken up into the clouds may have been how the eyewitnesses rationalized Jesus' departure, or it may have been presented to them in this way to help them understand that departure. He could just as easily have sunk down into the ground, but the optics of that would not be good. Or he could have taken one of the directions of the compass needle, but that might invite questions and hypotheses about the "meaning" underlying such a trajectory, whereas going "up" could only mean "to some place not here on earth." Or he could have just dematerialized before their eyes as he left our four-dimensional existence and entered the eleven-dimensional one that quantum physicists describe.

at the empty tomb may serve to link the resurrection of Jesus to Daniel's resurrection vision.

While many other pre-Christian civilizations have believed in some form of resurrection, it is the Christian religion which put a heavy emphasis upon that being a *physical, bodily* resurrection, rather than a resurrection of some kind of immaterial being:

> If the dead are not raised, then Christ has not been raised either. And if Christ has not been raised, your faith is futile; you are still in your sins. Then those also who have fallen asleep in Christ are lost. If only for this life we have hope in Christ, we are of all people most to be pitied.... If the dead are not raised, "Let us eat and drink, for tomorrow we die." (1 Cor 15:17-9, 32)

The fact that Paul says here "then Christ has not been raised either" indicates he is speaking about a physical death, since one would not think Paul was teaching that Christ was *spiritually* dead and then raised.

This heavy emphasis on bodily resurrection may be puzzling, given the context of the explosive growth of the early Christian church. That is, this new belief swelled within a Hellenistic Greek society, and many of the most prominent church fathers were Greek and/or thoroughly Greek-educated, yet many Greek philosophers held a decidedly negative view toward matter and the physical body and sought for the soul to be liberated from being bound to the body (see page 46). Judaism had a great deal in common with Christianity (the same Scriptures; the same history; many of the same core values), and had already been operating for centuries in the same Greek context, yet it never grew to the proportions that Christianity did. How do we explain this? Endsjø explores in great detail the idea that Greek society in general, contrary to its philosophical elite, did not hold such strong antipathy to physicality or to the human body. He shows that Greek mythology is full of examples of gods and humans dying and then being resurrected, and humans gaining immortality through dying. Endsjø also shows that Greek society greatly valued physical beauty, the Olympic Games, and the pursuit of medicine, all of which put a premium on maintaining or enhancing the body.[71] So although the hard-core philosophers in their society might have "preached" an animosity against the physical and the bodily, and might seem to have preferred the idea of the soul being liberated from the body over the idea of the body and soul being resurrected together, Endsjø argues that Greek society in general was quite comfortable with resurrection and immortality of the physical body.

---

71. Endsjø, *Greek Resurrection*, 1–104.

Endsjø also emphasizes that we today should not be too quick to claim that, *in the minds of those early Greek Christians*, Christ's resurrection was completely different from the resurrection of the gods and humans of their mythology. He claims that scholarly evidence points to the contrary, and so for one to make such a claim is simply "a theological stand."[72] However, just as we saw in the gospel accounts that Christ's resurrected body was not merely a resuscitated physical body, but rather something strangely different from the physical, Paul also seemed to envision something more than just a physical, bodily resurrection: he saw an entirely new existence within a "spiritual body," thus agreeing with Daniel (who envisioned a celestial existence, shining like the stars), and standing against 2 *Maccabees* (which envisioned simply a full restoration of the person's earthly body). His road-to-Damascus vision of Jesus speaking from a blinding light may have brought to Paul's mind Daniel's words of the resurrected believers shining "like the brightness of the heavens . . . like the stars" (Dan 12:3). In many of his writings, Paul refers to Christ being the "first to rise from the dead," the "first-fruit," and the "firstborn" (Acts 26:22–23; 1 Cor 15:20; Col 1:15, 18; Rom 8:29), implying that many others would follow suit.[73] This view is echoed by James (Jas 1:18) and the writer of Hebrews (Heb 2:10; 11:35). Paul's view of the resurrection body is more developed than Daniel's image of certain believers shining like the stars. He saw this resurrection being promised to *all* believers, and the physical body being transformed (Phil 3:20–21) into an entirely new kind of "spiritual body" (1 Cor 15:44), one which is glorious (Phil 3:20–21), imperishable (1 Cor 15:42, 50, 52–54), immortal (1 Cor 15:53; 2 Cor 5:4; 1 Thess 4:16), bears the image of the heavenly Man (Jesus Christ; 1 Cor 15:49), and within which "we will be with the Lord forever" (1 Thess 4:17). The basis for this view was not merely intellectual or even partisan, but was also experiential: he encountered, heard from, and fellowshipped with a resurrected Jesus Christ (Acts 22:6–10; 1 Cor 9:1; 15:8).

Once again, we need to address the fact that this new Christian reformulation of the concept of resurrection took place in a similar context as Daniel's recasting of it (see page 146). That is, the faithful in Daniel's era faced intense opposition and even martyrdom, and similar fierce opposition was ramped up once again in the first-century era (in fact, Paul was for a time an active agent in that anti-Christian persecution). Now Paul himself and his fellow believers faced martyrdom for *their* Christian belief (Acts 21:13; 2 Tim 4:6–8; Phil 2:17). Elledge contends that Paul, Daniel, and the

---

72. Endsjø, *Greek Resurrection*, 102.

73. Madigan and Levenson, *Resurrection*, 24–41; note: "firstborn" also has implications with respect to a superior status among siblings.

writers of 2 *Maccabees*, *On Resurrection*, and *Pseudo-Ezekiel* all framed this problem (of persecution and martyrdom) and its solution—the promise of resurrection, including full and complete restoration of ravaged bodies, celestial existence and rewards for the faithful, and judgment and fiery destruction for the unfaithful—in the context of theodicy.[74] They needed some kind of promise on which to hang their hope and obtain the courage to step boldly forward and offer everything—including their bodies and their lives—for their faith. However, the fact that this new revelation came at a time when it was most needed does not necessarily mean it was solely a human invention. The dramatic change in the disciples following the crucifixion is clear evidence that they saw something powerful and convincing. After deserting Jesus for fear of their life, denying having known him, and later hiding behind locked doors, they suddenly gained the boldness to confront the religious leaders and preach the gospel message of Jesus having been resurrected, at a time when those religious leaders could have easily researched the claim and demonstrated convincingly against it. This is strong evidence that their new view on resurrection was much more than just a coping mechanism or tool of propaganda. It could equally be seen to be a new revelation—part of the progressive revelation (see page 190)—from God for the church at a time when they would need it most.

Thus, "the Old and New Testament perspectives on resurrection are significantly different. In fact, for many scholars they are not just distinctive but actually contradictory."[75] Other scholars claim that glimpses of these ideas of resurrection and eternal life were always present in the Old Testament "in embryonic form," and were only discerned centuries later in the light of ongoing theological discussion and life's experiences (including the exile and the arrival of the Messiah): in other words, a form of hermeneutical spiral. For example, Routledge positions himself amidst the debate between those on the one hand who claim the Old Testament is largely silent on the matter of the afterlife and that the concept of the resurrection is a late borrowing from foreign influences by the writer of Daniel, and those on the other hand who claim to the contrary. Routledge sees "a growing consensus that Sheol is not to be seen as the final destiny of all the dead (though there is no clear view of an alternative final destination) and that final resurrection is not a foreign import, but the articulation of an idea already present, in embryo, in the earlier Old Testament faith."[76]

---

74. Elledge, "Resurrection of the Dead," 43.
75. Johnston, *Shades of Sheol*, 16.
76. Routledge, "Death and the Afterlife," 22.

Although many Christians today would see Paul's treatise on the resurrection as the final authority on the matter, many church fathers just one or two centuries after Paul did not hold to his views. In the second century, Tertullian used a more materialistic language than Paul (referring to "resurrection of the *flesh*," rather than that of the body) and Athenagoras of Athens even more so, while Origen in the third century CE took a more immaterialist view than Paul.[77] The Nicene Creed and Apostles' Creed, formulated in the fourth century CE, specifically refer to a belief in bodily resurrection, but fittingly do not provide any details on what is meant by that statement. It is worth pointing out, however, that an earlier version (second century) of the creed referred to *"the flesh"* being resurrected, but this was later changed to "the dead."[78]

Hades: Sheol and the Afterlife Become Hellenized

In the section above, we explored how Jewish thinking about resurrection evolved during and after the exile in Babylon and during the first few centuries of the early New Testament church. Here, we will look at how their thinking specifically about Sheol evolved during that same time period. For millennia prior to the exile in Babylon, the ancestors of the nation of Israel held to the idea of their dead residing in Sheol, which was more of a repository than a place of punishment (or of reward). It certainly was not a happy place to be, but was never associated in the preexilic texts with flames and torture. Neither do their older (preexilic) writings feature any other place of punishment (eternal or final).

This began to change when Israel settled in Canaan, nationalized, and set up Jerusalem as their most holy city. Immediately outside and to the south of Jerusalem lay the Valley of Hinnom, which became "the dump of that holy city, where fires were always kept burning to destroy the refuse."[79] This Valley of Hinnom came to be radically transformed in the minds of the exilic and postexilic Jews. In *Enoch* (v. 26), it becomes "the abode of the rebels against God, and that here 'in the last time [it] will serve as a drama of a righteous judgment before the righteous for all eternity.'"[80] During the reigns of Ahaz and Manasseh, the Israelites sacrificed their children in this valley to the Ammonite fire god and king of the dead—Molech—by literally

77. Elledge, "Resurrection of the Dead," 45; Madigan and Levinson, *Resurrection*, 231–33.
78. Moehlman, "Origin of the Apostles' Creed," 302.
79. Montgomery, "Holy City," 34.
80. Montgomery, "Holy City," 33.

throwing them into/through a furnace (2 Kgs 23:10; Jer 7:31; 32:35).[81] Jeremiah prophesied about God's judgment on the apostate Israelites, and that this same valley would become the "Valley of Slaughter" (Jer 7:32; 19:5–7). Isaiah prophesied of the destruction of the Assyrians by fire in a fiery furnace (Isa 31:8–9), and of a climactic slaughter of the wicked whose bodies would be burned by a fire that is not quenched and eaten by worms that do not die (Isa 66:24). This valley thereby became a metaphor of a place of divine judgment associated with fire, death, and bodily destruction, and came to be called Gê-hinnôm or Gehenna. Jesus himself referred to this place several times, warning his listeners to avoid being "thrown into the fire of Gehenna" which "never goes out" (Matt 5:22, 29–30; Mark 9:43, 45; Luke 12:5), where both body and soul could be destroyed (Matt 10:28) and the condemned can never escape (Matt 23:15, 33). It is clear that he was speaking metaphorically: there is no reason to think that this valley just outside the city of Jerusalem is the location of a literal cosmic hell for all of humanity. James, the brother of Jesus, also refers to the tongue being set on fire by Gehenna itself (Jas 3:6), and also referred to a "Day of Slaughter" (Jas 5:5). On the other hand, it is noteworthy that Paul, who wrote the majority of the New Testament books, did *not* refer to Gehenna (although he does write about judgment, death, and destruction).

Many scholars have contemplated how this valley became such a metaphor for hell, and some found a possible influence from Babylonian mythology. A common motif in the ancient Near East was "the holy mountain" of the local deity,[82] and Jewish literature makes numerous references to holy mountains: many key events in the history of the nation of Israel occur at Mount Horeb or at Mount Sinai, not least of which include Moses at the burning bush (Exod 3:1—4:17) and Israel receiving the Law (Exod 19 and 20).[83] More germane to this discussion, though, there are also numerous references to Mount Zion being the holy mountain of YHWH himself (for example, Ps 48), and Jerusalem, the holy city of YHWH, being built upon it.[84] Israel would have been familiar with Ekur, the mountain house of Enlil, the chief god of the Babylonians, particularly when they were taken into captivity by Babylon following the collapse of their own nation and religious system. During that period, many of the Old Testament texts were either composed and/or markedly edited, and there was considerable development

---

81. Montgomery, "Holy City," 34; Zahnd, *Sinners in the Hands*, 123; Tate, *Biblical Interpretation*, 40; Scharen, "Gehenna," 327.

82. Noort, "Gan-Eden," 27.

83. It is disputed whether these two mountains are distinct or one and the same.

84. Montgomery, "Holy City," 27.

at that time of the religious concepts of Sheol and resurrection that was undoubtedly influenced by their cultural milieu (as already discussed above). Ekur was the center of the Babylonian cosmos, the conjunction between heaven and earth. It also featured a temple from which Enlil issued his decrees and divine laws. And it was a place of judgment, including a dungeon at the base of their holy mountain believed to be connected to the netherworld—*Aralû* or "House of Lament," the Babylonian equivalent of Sheol—where the damned were sent after judgment. Likewise, the ancient Greeks mythologized about Tartaros, a place of punishment for certain individuals (divine or human) who were exceptionally and irredeemably evil (see page 38).[85] The writers and editors of the Old Testament texts would certainly have been exposed to this Babylonian and Greek imagery: it is a matter of conjecture whether/how it influenced their writing of the Old Testament texts as they spiritualized their eschatology. Thus, together with their new Hellenized version of resurrection and the afterlife, the Hebrew concept of Sheol was transformed into a metaphorical Valley of Slaughter and destruction (Gehenna) and then into the literal Greek concept of Hades (which we in the modern era now refer to as "hell").[86]

The Garden, Eden, Paradise, and Heaven: Concepts in Flux

In the same way that Old Testament texts dating to before Second Temple Judaism do not clearly mention a postmortem place of punishment, they equally do not *clearly* describe one of reward for the faithful and righteous. As a concept in the Bible, Eden/paradise is an ever-moving target, vacillating back and forth in space, time, and description.

In terms of location, the garden in Eden is first presented as being somewhere in Mesopotamia, but later the land of Eden is linked to a Canaanite tribal group (2 Kgs 19:12; 2 Chr 29:12; Amos 1:5), and in other passages we find Eden in heaven (Rev 2:7; 22:1–4). (To be precise, the Eden described by John in Revelation is not the same one as that described by the author[s] of Genesis). The apocryphal *Book of Enoch* also refers to a paradise in heaven, and this is further developed later yet in rabbinical literature.[87]

In temporal terms: is the garden from our long distant past (the temporary home of Adam and Eve, but now closed to all humans by cherubim

---

85. Scharen, "Gehenna," 327.

86. Harris, "Meaning of the Word," 129.

87. Landman, *Universal Jewish Encyclopedia*, 515–20; Tigchelaar, "Eden and Paradise," 37–62.

bearing a flaming sword; Gen 3:24); is it in the author's present but still our past (the references to Canaan); is it an ongoing present state or place (to which Enoch and Elijah were taken, and the promise made by Jesus to the thief on the cross; Luke 23:43); or is it in the future (the final destination for all believers *after* the New Jerusalem is ushered in)? (This problem is solved by proposing it to be an eternal place, although again the paradise / new heaven presented in Revelation is not the same as the garden created by YHWH in Genesis.)

In form and structure: is it a garden with a gentle flowing stream (Gen 2:8–14), a forest (Ezek 31:8–9, 16, 18), a mountain (Ezek 28:14, 16), or the heart of a cosmopolitan city (Rev 22:1–5)? Genesis presents it as a garden small enough for one human couple to take care of, but Revelation describes it as something which "all the nations" can inhabit. Is it a physical place, a mythological one, or a spiritual one? We will address these questions and many others like it in the sections below.

The Bible is similarly unclear about heaven. Genesis implies "heaven" is the third sphere surrounding the newly created cosmos—the "place" where God and the angels dwell, although YHWH is otherwise seen to be unconstrained by space or time—while apocryphal literature refers to seven or ten heavens. In the third of the seven heavens, or the seventh of the ten heavens, one would find heaven's paradise containing the treasures of life and of righteousness for the soul.[88] The numbers three, seven, and ten were employed purely for philosophical/theological reasons—they were arbitrarily declared to be perfect or divine numbers—rather than as an observable or demonstrable fact. How could one ascertain or even "prove" such numerical details as the existence of ten heavens?

Once again, all of this lack of clarity owes in large part to the fact that these concepts of the afterlife were still evolving in the minds of the biblical authors, as we will see in the next few sections below.

The Garden of Eden in the Old Testament

The second chapter of Genesis introduces the readers to *gn-'dn*, or *Gan Eden*, which is translated in English as the garden of Eden. Waltke explains that the Hebrew root *gnn* "probably denotes an enclosed, protected area where the flora flourishes."[89] *Gan Eden* is presented as a garden small enough to be tended by two individuals—Adam and Eve—who enjoy an intimate relationship with God: they walk with him in the cool of the day

---

88. Landman, *Universal Jewish Encyclopedia*, 298.
89. Waltke, *Genesis*, 85.

(3:8). It featured all kinds of "trees that were pleasing to the eye and good for food" (2:9), and in the middle of it were found two unique trees with particular theological relevance: the Tree of Life and Tree of the Knowledge of Good and Evil (2:9). It also featured "all the wild animals and all the birds in the sky" that God had made (2:19). Finally, it was indeed an enclosed space in the sense that its occupants could be driven out and prevented from reentering by "cherubim and a flaming sword" (3:24).

The book of Genesis defines a fairly precise location for *Gan Eden*: a river running through it becomes the headwaters for four rivers: the Tigris, Euphrates, Gihon, and Pishon rivers. Two of these—the Tigris and Euphrates—still exist even today, being located in the Mesopotamian basin region: in fact, these two rivers connect in the region of Sumer. That latter geographical detail is particularly noteworthy: many scholars have commented that the first eleven chapters of Genesis bear remarkable similarity to ancient Sumerian mythological literature (as explained above on pages 14–20). This frequently leads to hypotheses of the ancient Hebrew authors of Scripture interacting closely with these Babylonian narratives and the Babylonian zeitgeist: not necessarily outright borrowing, but at the very least reacting to them in polemic fashion (see pages 21–23).[90]

The third river—the Gihon—is said in this passage to "wind through the entire land of Cush" (Gen 2:13), otherwise known as Ethiopia. A much later passage refers to the Gihon as a spring near Jerusalem (2 Chr 32:30), which flows eastward toward the Kidron Valley and watered the gardens and parks planted by the kings of the Davidic dynasty,[91] but this spring is hardly big enough to encircle "the entire land of Cush."[92] Both geographical locations are quite distant from Sumer, and their waterways have never been known to connect to the Euphrates or Tigris Rivers.

Finally, the Pishon River does not exist today, and even the first-century Jewish historian/philosopher Philo was not able to specifically identify it with any waterway existing in his era.[93] It is said to "wind around the entire land of Havilah," that land being mythologized as a source of gem stones (thought by some to have magical powers) which are carried by the Pishon to the shores of the Red Sea."[94]

---

90. Clouser, "Reading Genesis," 237–61; Pinnock, "Climbing Out," 143–55; Harlow, "Creation according to Genesis," 163–98; Fugle, *Laying Down Arms*, 241; Sanders, *Invention of Hebrew*, 76; Watts, "Making Sense," 5–12.
91. Madigan and Levenson, *Resurrection*, 85.
92. Noort, *Gan-Eden*, 29.
93. Niehoff, "Philo's Scholarly Inquiries," 40–41.
94. Bauckham, "Paradise," 46.

The association of paradise with the two rivers which can be precisely located on modern maps (the Tigris and Euphrates) has long encouraged speculation and even exploration—by notables including Christopher Columbus and Amerigo Vespucci—for the precise location of a physical paradise on earth.[95] Of course, they never did find the object of their search. What does the modern exegete do about the other two rivers which cannot be so precisely located? Some ancients, including Josephus,[96] identified the Pishon with the Nile (because the land of Havilah is "opposite Egypt in the direction of Assyria" [Gen 25:18; also see 1 Sam 15:7]), and the Gihon with the Ganges River, even though neither the Nile nor the Ganges are anywhere near the Euphrates or Tigris rivers.[97] On the other hand, rather than trying to identify each of the four rivers with some actual correlate on earth, could the author of Genesis be referring to two literal, physical rivers and two fictional, legendary ones (one of which is the source of magical stones) as a literary tool to paint an image within the minds of an ancient Hebrew of *Gan Eden* being a surreal place that bridges the world of the real/earthly and the world of the mythical/heavenly?

The authors of Genesis take the trouble to further associate *Gan Eden* with an excellent place to find aromatic resin, onyx, and gold (in fact, "the gold of that land is good"; 2:12), which always struck me as a red herring in this ancient passage. Bauckham unpacks in detail this puzzling point-of-fact by exploring several passages in *1 Enoch* which refer to collections of gem stones from the land of Havilah: twelve stones that were set on the priestly ephod opposite the twelve that Moses set on the breastplate (both garments being then placed in the ark of the covenant; Exod 28:15–30), as well as seven stones that decorate seven golden idols (of sacred virgins) of the Amorites. Later, Ezekiel describes a king of Tyre who "[was] in Eden, the garden of God" and who was adorned with various gem stones (28:12–19): the Masoretic Text of this passage lists nine of the twelve that were also set in the breastplate fashioned by Moses, while the Septuagint lists all twelve. Onyx and gold are mentioned in all of these passages. With this additional information on the table, this odd mentioning of the treasures to be found in Eden now seems to serve a literary, even theological, purpose: it links this quasi-mythical garden of Eden with a priestly function on earth. Walton has hypothesized that the first chapter of Genesis is about the inauguration of a

---

95. Bockmuehl, "Locating Paradise," 192–209; Auffarth, "Paradise," 172; Scafi, *Maps of Paradise*, 78, 80.

96. Josephus, *Antiquities*, 1:38–39; Noort, "Gan-Eden," 29–30.

97. Bockmuehl, "Locating Paradise," 198–99; Noort, "Gan-Eden," 29–33; Auffarth, "Paradise," 172.

cosmic temple (a place for those priests to serve) rather than about material origins (see "Theological Significance of the Garden," below).[98]

Two other passages in the Hebrew Bible are closely linked to the second and third chapters of Genesis and a reconceptualization of the garden and the fall which occurred there: both of these describe the fall from grace of an individual who "reached too far" and was punished for doing so. Returning to Ezekiel chapter 28, written long after the *Gan Eden* of Genesis was introduced, we find a king of Tyre who, just like Adam in Genesis, is said to have been in a garden named Eden, claimed wisdom and knowledge that he believed made him like a god, and was therefore confronted by sword-wielding opponents. In fact, Ezekiel refers to fire and flames (possible allusions to the cherubim guarding reentry into Eden, in Gen 3:24?) and the king of Tyre being expelled from Eden, later to be burned to ashes (returned to dust?).[99] However, there are important differences between Genesis 3 and Ezekiel 28.[100] Genesis's *Gan Eden* is set in a small well-watered garden in which God and his two newly created humans enjoy close fellowship, while Ezekiel's "*Eden, Gan Elohim*" (28:13) includes the mountain of God, "the cosmic capital from which the deity exercises sovereignty over his universal domain," and which may have had a volcanic aspect to it, given that it featured "fiery stones" (28:14, 16).[101] Adam is portrayed as a naïve and childlike human (lacking knowledge of good and evil) who tills the earth from which he was made, while the king of Tyre is portrayed as a cherub "full of wisdom" (Ezek 28:12) who dwells on God's holy mountain from which he was thrown down to the earth. Others, however, interpret that passage to refer to Lucifer, largely on the basis of what Jesus said about seeing Lucifer fall like lightning from heaven (Luke 10:18) even though his words here more closely parallel the fall from heaven of "the morning star, son of the dawn" described by Isaiah (14:12).[102]

A later passage in Ezekiel—ch. 31, a prophecy against Pharaoh king of Egypt—also refers to *Gan Elohim*. This passage makes numerous references to the trees of Eden: cedars, junipers, and plane trees (31:8). However, this passage is clearly very symbolic in that it refers to "Assyria, once a cedar in Lebanon" (31:3) being higher than all the other trees of the field (31:5), with birds nesting in its boughs and animals giving birth under its

---

98. Walton, *Lost World of Genesis One*, 86, 92; also see Waltke, *Genesis*, 81.
99. Tigchelaar, "Eden and Paradise," 27.
100. Noort, *Gan-Eden*, 27.
101. Madigan and Levenson, *Resurrection* 82–83; also see Noort, *Gan-Eden*, 27. Some think these refer to members of the Divine Council: see 1 Enoch 17:1–6.
102. Noort, *Gan-Eden*, 22.

branches (31:6), all of which are later interpreted to be the nations of the earth (31:6, 12). (Interestingly, paralleling the fate of the king of Tyre in Ezek 28 and the primal pair in Gen 3, this "tree" is then punished for its pride by being cut down [31:10–2].) With this literary context in place, the cedars, junipers, and plane trees found in *Gan Elohim* would appear to be metaphors for nations or people groups: they are said to be unable to rival nor compare to this Assyrian nation/king who is the "cedar of Lebanon" (31:8), and had also "like the great cedar, gone down to the realm of the dead, to those killed by the sword, along with the armed men who lived in its shade among the nations" (31:17). Even though all these "trees" therefore appear to represent various persons or people groups, this passage nonetheless makes clear that Ezekiel imagined the garden of God being like a grand well-watered forest harboring tremendous numbers of wild animals and birds. This imagery closely parallels that in Genesis, which describes a garden containing all kinds of birds, wild animals, and trees, with two trees in particular meriting special mention.

The garden of Eden is referred to in other passages of the Old Testament, but only as a comparator or ideal, somewhat in a sense like someone today might describe the Rock and Roll Hall of Fame as "heaven." When Abram gives him the choice of land, Lot surveys and chooses the well-watered Plain of Jordan toward Zoar, which was said to be "*like* the garden of Eden" (Gen 13:10; emphasis added). The prophet Joel foretells of the destruction of Zion by an army of locusts: "Before them the land is *like* the garden of Eden, behind them, a desert waste" (Joel 2:3; emphasis added). The prophets Isaiah and Ezekiel foretell the opposite when Zion is restored following her captivity: "He will make her deserts *like* Eden, her wastelands *like* the garden of the Lord" (Isa 51:3; emphasis added) and "This land that was laid waste has become *like* the garden of Eden" (Ezek 36:35; emphasis added).

The Garden of Eden in the Hellenic Period

In the Hellenized minds of believers in the few centuries before Christ, the Old Testament Hebrew concept of *Gan Eden* became connected to the Hellenic Greek concept of paradise. Bremmer and Scafi both provide excellent reviews of the origin and evolution of the word "paradise."[103] The Medes of the early Achaemenid Empire (sixth century BCE) maintained vineyards and small orchards within walled enclosures, in which they would

---

103. Bremmer, "Paradise in the Septuagint," 1–20; Scafi, *Maps of Paradise*, 9–11; also see Stroumsa, "Paradise Chronotrope," 1; Charlesworth, "Origin and Development," 222.

THE AFTERLIFE 163

also store food supplies, sometimes including domesticated animals such as sheep. These walled enclosures were referred to using various combinations and derivations of their words for "around" (*pari*) and "wall" (*daeza*). As the Achaemenid Empire grew in power, these small *paridaeza* grew in size until they became tree parks or vast orchards filled with wild animals large enough to entertain hunting parties on horseback. When Alexander the Great overthrew the Persian empire in 331 BCE, the "paradises" which were confiscated by the Greeks—now called *paradeisos*—became less like hunting parks and more like vast "tree gardens" full of vegetation and flowing streams of water.[104] As time went on and the Romans overthrew the Greeks, these gardens became more lush, landscaped, manicured, and filled with exotic animals like swans, parrots, and peacocks.

Bremmer goes on to explore why the writers of the Septuagint—the Greek translation of the Hebrew Bible, our "Old Testament"—chose to render the Hebrew expression *gn-'dn* using the Greek term *paradeisos* rather than *kêpos*, the Greek word for garden.[105] His conclusion is that the *kêpos* was generally smaller, simpler, and more associated with residential housing, making/drinking alcohol, and even with sexual activity, while the *paradeisos* was massive, lavish, and associated with royalty. In his words: "For the Jewish translators the word *kêpos* will have hardly conjured up the image of a royal park worthy of Jahweh [sic]."[106]

Hellenized Jews Relocate Paradise to Heaven

The New Testament makes no overt references to the garden of Eden (there is a vague allusion to it in Rev 22), and has only three references to paradise: in Jesus' promise to the thief on the cross (Luke 23:43), in Paul's vision of being caught up to the third heaven (2 Cor 12:2-4), and in John's vision of the New Jerusalem (Rev 22:1-4; also see Rev 2:7). All three imply paradise is a destination for the righteous in the afterlife, even if only a temporary one. Other references can be found in noncanonical literature dating to around the same period. While Genesis (5:24) and the book of Hebrews (11:5) mention cryptically that Enoch avoided death by being "taken away" by YHWH, the book of *Jubilees* fills in some of the details, claiming he was carried alive by angels to the garden of Eden, located somewhere between earth and heaven (*Jub.* 4.23). Similarly, in the *Testament of Abraham*, God

104. Stroumsa, "Paradise Chronotrope," 1, 3; Schaper, "Messiah in the Garden," 19; Bremmer, "Paradise," 1; Scafi, *Maps of Paradise*, 10.
105. Bremmer, "Paradise, from Persia," 17-19.
106. Bremmer, "Paradise, from Persia," 17-18.

says: "Take my friend Abraham to Paradise, where are the tents of the righteous.... There is no toil there, no grief, no sighing, but peace and rejoicing, and endless life" (*Testament of Abraham* 20:18–19). Finally, other texts hold this promise out to believers in general (*2 Enoch* 65:6, 10; *Testament of Dan* 5:12). However, scholars question whether this idea of Eden as the final destination for all righteous believers was well-developed or widely accepted at the time of Christ, given that it is mentioned in this way only three times in the New Testament, and is not found in other texts where one should expect to find it if this idea were indeed commonplace (for example, it is not mentioned in Josephus's *Jewish Antiquities*).[107] The idea became more common in second Temple Judaism and later Christianity (*4 Ezra*; *2 Baruch*; *Biblical Antiquities of Pseudo-Philo*).

In Paul's reference to paradise, he claims to have been caught *up* to the third heaven and then caught *in* to paradise (2 Cor 12:2, 4), leading some to wonder whether this was a two-stepped journey and that paradise is a specific area within heaven,[108] much like the author(s) of Genesis describe the primeval garden as being a specific area *in* Eden, not Eden itself (Gen 2:8), and John describes the garden as a specific area within the New Jerusalem rather than as all of heaven itself (see Rev 2:7 and Rev 22:1–2). The uncertainty is clouded further by Paul not explaining what he himself means by "the third heaven." Some point to *2 Enoch* 8:1–8, in which Enoch is carried up to the third of seven heavens (a longer version of *2 Enoch* refers to ten heavens, though most agree that this wording was not part of the original text) where paradise is found (the divine presence residing in the uppermost of the seven or ten heavens). Others point to Genesis chapter 1, in which God creates a three-layered cosmos by inserting a firmament or vault (which he named the "sky") to separate "the waters above" from "the waters below," and then taking residence in a third level beyond these waters.

Later Christianity Reformulates Paradise Further Yet

The concepts of paradise and heaven continued to be developed within rabbinic Judaism and in early Christianity. For example, *2 Enoch* describes paradise as being divided up into forty parts. More importantly, however, there is a growing conviction that all of God's righteous followers can find their ultimate destiny in heaven (this contrasts with the thinking of the ancient Hebrews: for them, there was no clearly articulated separation of the righteous and the wicked in Sheol, nor was there an overt promise of eternal

---

107. Goodman, "Paradise, Gardens," 57–63; Macaskill, "Paradise," 64.
108. Macaskill, "Paradise," 67–71.

life until the time of Daniel, and even that was indicated only for *some* of the righteous [12:1-3]). However, a lack of unanimity grew with respect to the means by which one gets to heaven: the relative roles played by Christ who makes that possible, by the believer taking certain steps in this life, and possibly through time spent in Purgatory in the afterlife. But that is atonement theology, a completely different topic than the nature of the afterlife: we will not touch upon that here.

Benjamins provides an excellent review of how paradise was understood in the early church, including the fact that over time, access to paradise is increasingly connected to a life of deprivation and suffering: first in the form of virginity and sexual chastity, but later also to martyrdom and the monastic life.[109] Many early leaders—including Ireneaus, Tertullian, and Cyril of Jerusalem—taught that paradise was now available to all believers.[110] Auffarth extends that view into the Middle Ages,[111] describing a growing separation between the men and women of the church against the laity, a separation which played into an emerging conceptual difference between a present paradise and a future paradise. That is, although it was clear that there was a New Jerusalem or new heaven—which included a kind of spiritual form of paradise—to be entered into *following the Final Judgment*, it also seemed that there was a kind of paradise which could be experienced in the here and now, the same one to which Enoch and Elijah were translated, and presumably the one which the thief on the cross visited with Jesus. The belief grew that it was possible for believers to immediately access that present paradise prior to our final graduation to the spiritual paradise in heaven. It was pointed out that monks and nuns already lived a life of chaste virginity, meditation, and tilling the soil to produce the food they eat: in other words, the same kind of life lived by Adam and Eve in the historical paradise on earth. Common people, on the other hand, forsook that paradisiacal life by participating in sexual reproduction and all manner of earthly business. In Auffarth's words: "Unlike the Muslim paradise, the Christian paradise was not open to all; as a place of final rest it was restricted to but a few holy men and women. . . . But in the course of the eleventh century there was a change: the laity demanded its own path to salvation."[112] One solution to this conflict was to extend entry to those who participated in the Crusades: "Killing an enemy of Christ was to [Bernard of Clairvaux]

---

109. Benjamins, "Paradisiacal Life," 155-56.

110. Irenaeus, *Adversus Haereses*, V.6.1; Tertulian, *Adversus Marcionem*, II.IV; Cyril, *Catechesis*, XIX.9.

111. Auffarth, "Paradise Now," 168-79.

112. Auffarth, "Paradise Now," 173.

an act of salvation."[113] Others sought to gain entrance by practicing various forms of mysticism and an ascetic life.

The Renaissance took the shine off the concept of paradise, at least the idea of a physical historical one. Up till that point, world maps were oriented in such a way that paradise was put at the top of the map and Jerusalem at the center, solely for theological reasons.[114] However, scientific advances such as the invention of the compass and the building of ships capable of sailing across oceans led to the redrawing of those maps, in part by now reorienting them with magnetic north at the top. In addition, explorers began to massively redefine our understanding of the geography of earth: in the process, paradise shrank down in size and was increasingly accompanied by question marks as the maps became more detailed and accurate. It goes without saying what eventually happened to the concept of paradise as a physical place on earth. That reconceptualization extended into other categories of thinking such that paradise increasingly became little more than a metaphor. The Renaissance, however, did not greatly impact the Christian concept of "heaven," nor did other later movements such as the Scientific Revolution. That concept persisted, and has also continued to evolve. But the conceptualization that many Christians have today is not radically different from the one held even a thousand years ago: only variations on a common theme.

Theological Significance of the Garden

The story of Adam and Eve in the Edenic garden is central to so much of Christian theology. Connected to it are large theological issues as wide-ranging as the fall, "original sin," death, our need for a Savior, restoration, redemption, and our eternal hope, as well as less significant issues such as weeds in our gardens, our fear of snakes, why the latter have no legs, and pain in childbirth. And yet, despite this broad and deep significance of *Gan Eden*, the picture of those two individuals in the garden is only ever drawn *once* in the Bible (here, in the first few chapters of Genesis). Outside of Genesis, the Old Testament mentions Adam once, but not with any connection to the garden or to Eve (he is simply listed as the first in an extended genealogy; 1 Chr 1:1). Neither Eve nor the story of the temptation and fall are ever referred to in biblical texts outside of the two chapters in Genesis. Other books in the Hebrew Bible mention the garden several times, but never together with Adam and Eve. Jesus does not explicitly mention this vignette, even though

113. Auffarth, "Paradise Now," 177.
114. Scafi, *Maps of Paradise*, 32.

some would say this was the primary target of his mission, and neither does any New Testament author mention it other than Paul.

YHWH's garden in the land of Eden is presented as the first temporary home of the first human couple (Gen 2), but it is never portrayed in the Old Testament as a postmortem destination for righteous believers in general. In fact, the land of Eden later became the ancestral home for another pagan nation (2 Kgs 19:12; 2 Chr 29:12; Amos 1:5). The Old Testament does refer to Sheol as a postmortem resting place—for the righteous as well as for the wicked—but Sheol is never described as a lush, well-watered garden with trees and animals, nor as the mountain of God.

Walton interprets the first chapter of Genesis to be describing the creation of a cosmic temple by YHWH, rather than as an ancient "scientific" narration of material origins.[115] The culmination of this temple-building would be the installation of an icon of the God to whom the temple is dedicated: in the cosmic temple of Genesis 1, humans are that *imago Dei*. However, Walton does not expend much effort toward folding the second chapter of Genesis into this exegesis: for example, by positing the small *Gan Eden* as some particularly sanctified portion of that relatively larger cosmic temple, such as a kind of holy of holies.

Jesus and "the Kingdom of Heaven"

Did Jesus teach anything about heaven? Some readers may be quick to reply in the affirmative, pointing to the numerous passages in which Jesus says "the kingdom of heaven is like . . ." But do these really give us a picture of heaven? For two reasons, I do not think the answer to this question is so straightforward: most readers will likely already acknowledge the first of these two, but the second may surprise them.

First, we need to recognize that Christ was often using metaphor when he said "the kingdom of heaven is like . . ." He did *not* teach that heaven *is* a mustard seed, but rather that the object of his statement is *like* a tiny mustard seed that grows into something incomprehensibly bigger, something that can dominate the garden and accommodate many birds (Matt 13:31). He did *not* teach that heaven is a handful of bacteria (yeast), but that we should be thinking of something as being *like* a very small—even microscopic—living thing that has the power to expand inanimate dough into something far bigger: in fact, big enough to feed a whole family of people (Matt 13:33). These two passages and many others convey the idea that the kingdom of heaven starts off small and grows organically to massive proportions. Other

---

115. Walton, *Lost World of Genesis One*, 86, 92; also see Waltke, *Genesis*, 81, 101.

metaphorical passages teach that the kingdom of heaven is worth pursuing at all costs (Matt 13:44, 45), or that it produces all kinds of valuable things (Matt 13:47, 52). Christ said many diverse things about "the kingdom of heaven," and it is important for the reader to always note the key phrase "is like" and to see what follows as metaphor. The appropriate response to that realization is to avoid the superficial or lazy approach of simply reading that passage at face value, but to instead do the harder work of extracting the actual meaning: to read it literarily, rather than literally. Metaphors and allegories are intended to convey some aspect of truth, but rarely capture the entirety of that truth: they always fall short or fall apart when a critical listener begins to engage with it and challenge it too literally. Some may be quick to deny that they themselves lean toward such superficial readings, but fail to realize they do exactly this when it comes to discussions about hell (for example, some see hell as a place of flames and smoke because this imagery is explicitly employed, albeit *in metaphors*).

The second point is more subtle and easily missed. When he employed these metaphors, Jesus never said "heaven is like . . . ," but rather "the *kingdom* of heaven is like . . ." The word "heaven" is being used as an adjective, not the object of the sentence. This is an important distinction. The "*kingdom* of heaven" (or "the *kingdom* of God") does not need to refer to some distant place in the sky, let alone to a place at all. Kingdom refers to a system of government, a hierarchy of ruling, an economy. Christ taught us to pray "thy kingdom come . . . on earth as it is in heaven": clearly "kingdom" and "heaven" are two different things here (earth could not possibly accommodate heaven, even if one ignored for the moment the category error of putting material earth and a spiritual heaven on the same level). He began his ministry by announcing that "the kingdom of God has come near" (Mark 1:15): a *system* can be brought to a tiny rock orbiting one of uncountable stars in the universe, but not heaven itself. He taught that "from the days of John the Baptist until now, the kingdom of heaven has been subjected to violence, and violent *people* have been raiding it" (Matt 11:12; emphasis added): how could this be referring to a spiritual place far removed from earth, and presumably operating outside of the dimensions of human activity, let alone be vulnerable to human attack? He also said that "in fact, the kingdom of God is among you" (Luke 17:20–21; some translations have this as "in you"): it would be pretty hard to get people living during the Roman occupation of the first century to agree that heaven itself was among or in them.

Christ's references to "kingdom" bring us directly to the heart of his message: the gospel. That word "gospel," or its equivalent, will not be found anywhere in the Old Testament: there are hints of it in various places, and one must read between the lines to find them. But then it suddenly makes its first of dozens of appearances in the New Testament: mostly in the four historical accounts of Matthew, Mark, Luke and John, which are otherwise referred to as the "Gospel accounts." Admittedly, this is in part an artifact of the reality that the word "gospel" is derived from the Greek (see next paragraph), while most the Old Testament was written in Hebrew (some in Aramaic). But more than that, the word "gospel" in the New Testament refers to a very specific thing—a message—which is only vaguely alluded to in the Old Testament but brought out into the open in the New Testament.

It is easy for us today within contemporary Christianity to know what "the gospel" means (this would be yet another example of us retrojecting contemporary ideas or interpretations upon a first-century concept). Or at least we might *think* we know what it means. While speaking to a class of divinity school students, I began my presentation with a surprise quiz. The exam sheet said they had sixty seconds to give two definitions: first, "the American Dream," and second, "the Gospel." I had previously gone through a similar exercise with approximately thirty long-standing churchgoers not studying at the divinity school. Everyone found it easy to define the American Dream as something having to do with finding freedom, prosperity, and success. But the second question caused everyone to pause and really have to think how they would actually articulate their own definition. A few of the second group I tested simply said, "I don't know," even though they had been attending church for decades. Approximately a third of this lay group answered along the lines of "the gospel is the life-story/biography of Jesus" (in other words, the gospel stories themselves are the gospel message). All of the rest of those taking my surprise quiz—the divinity college student group and the lay group—gave a wide-ranging set of theological statements:

- "God is no longer mad at us. Jesus paid our sins, opened a pathway for us to God . . . we are now sons of God."
- "The fulfillment of the Old Testament."
- "Loving God and loving others as you would want to be loved."
- "Something in the world is broken . . . it can be fixed . . . Jesus is the solution."

My goal in conducting this exercise with them was to show and emphasize how poorly the church has equipped believers with a crystal clear, pithy statement of our primary mission. Try this with your own church group(s) to see if they fare any differently in being able to define the Gospel as quickly and uniformly as they define the American Dream.

Jesus commands his disciples to "go into all the world and preach the gospel to all creation" (Mark 16:15). Do you think those disciples started looking around at each other with puzzled faces, mouthing the words "What's the gospel?" and shrugging their shoulders in response? I do not. They had already encountered this word "gospel" many times long before they had met Jesus, and when they heard him reappropriate the word in an entirely new context, they knew exactly what he meant. The word that Jesus is quoted as saying (we do not know whether he said this in Greek or in his native Jewish Palestinian Aramaic tongue) is the Greek word εὐαγγέλιον (*euangelion*), which literally means "good news." That eventually became translated into the old English literal equivalent of "good news" or "good story"—*gōd spel*—and then was modernized into today's English word "gospel."

The *euangelion* was prevalent in first-century Hebrew society, but not in any religious sense: it was a political word borrowed by Jesus and his disciples from their surrounding Greco-Roman culture. There are numerous appearances of the *euangelion* in Greco-Roman literature in the century preceding the birth of Christ. During this time, the Romans ruled the world, but their Republic was being torn apart by internal conflict, corruption, violence, infighting between factions and self-appointed dictators. And that spilled out into their conquered regions, including Greece. The Greeks longed for peace, order, stability, and intellectual pursuit, but their view of the future was rather dim because of what was going on in the ruling government of their day . . . the Roman Empire. Then along came Caesar Augustus who took the role of emperor, did away with the Republican model of government, and worked to bring those values that the Greeks delighted to see in a government. These Greeks (and the Romans) had much to say in favor of this new Caesar, but one particularly relevant instance can be found in a letter from the Proconsul Paullus Fabius Maximus, dated several years before the birth of Christ, and decades before Christ began his ministry, copies of which have been found in several Greco-Roman cities. Central to my point is this excerpt from his letter (in which he quotes from a Greek high priest, who in turn is referring to an impersonal, universal life force):

> Providence, which has ordered all things and is deeply interested in our life, has set in most perfect order by giving us Augustus, whom she filled with virtue that he might benefit humankind, sending him as a savior, both for us and for our descendants, that he might end war and arrange all things. And since he, Caesar, by his appearance . . . surpassing all previous benefactors, and not even leaving to posterity any hope of surpassing what he has done, and since the birthday of the god Augustus was the

beginning of the good tidings [*euangelion*] for the world that came by reason of him.[116]

Decades before Christ began preaching his gospel and commanding his disciples to do the same, the Romans and Greeks were preaching their own *euangelion*. That good news was that a savior had been sent to the world to bring peace to all mankind: Caesar Augustus, who finally put an end to the internal conflict within the emerging Roman Empire, as well as the overt conflict it was waging with its conquered regions, and brought peace to everyone. That world peace—the *pax Romana*—brought stability and liberty from internal political turmoil, pirates on the seas, bandits on the highways, crooks in the cities, and from the barbarian hordes outside of the Roman Empire. It brought strong leadership (the catchphrase of the day was "Caesar is Lord"), and unity in the form of one common language, one common currency, and one common law. It brought freedom, or at least a form of freedom: even slaves had a degree of freedom that they did not have under the self-destructive Roman world before Caesar, and of course the Jews especially enjoyed their freedom to practice their Jewish faith. It brought the benefits of a modern civilization: technology, medicine, education, justice, and law. And it brought prosperity because of the common currency, the system of roads that the Romans built and the safety they ensured for the transportation of goods. This was good news indeed, and the Greeks and Romans loudly proclaimed it: "Caesar is Lord!"

Jesus appropriated that Greek word from Roman political propaganda: he claimed that he too had an *euangelion*, one that looked very similar to the Roman one. Christ also brought good news that likewise promised peace and unity . . . although this time between God and humans, and between different groups of humans, as is captured in the verse that is often quoted at Christmas time: "Peace on earth and goodwill to all men" (Luke 2:14). Just like the Roman gospel, Christ's gospel also brought liberty, although now from a very different form of captivity. And it brought its own equivalents of the benefits promised by the new Roman world system: instead of medicine, Christ's gospel brought healing (physical, emotional, spiritual), and instead of justice and law, Christ's gospel brought forgiveness and justification. Not only were there parallels in the promises and benefits of these two gospels, but those two world systems both came with exceptional costs. The Roman Gospel was forced on everyone, often through violence and ruthless punishment, sometimes to the point of persecution; it came with many rules and laws, exorbitant taxes, required exclusive worship of the emperor, and imposed a pagan Roman presence, which was at times offensive to the Jews. Christ's

---

116. Byrd, *Gospel of the Lord*, 8.

gospel was "enforced" by love—one was free to choose or reject it—but it required total surrender and exclusive worship of God.

It is worth adding that the Roman Empire used many other words in its propaganda which Paul and Luke later co-opted in their own writings.[117] The Greek word for "Lord"—*kyrios*—was one of the imperial titles used at the time for Caesar, and of course was later reappropriated by the followers of Christ. The phrase "Son of God," which readers will recognize as referring to Jesus, was long before used by those in the Roman Empire to refer to a "divinized" Caesar, son of Zeus. *Parousia* meant an "arrival" or "coming" of a high dignitary in the Roman Empire, and is used in the New Testament to refer to Christ's second coming. Likewise, the Greek *ekklesia* or "assembly" became the Christian word for "church." There are many other parallel usages of words by the Roman Empire and Christianity, to which the latter added a spiritual element.

It is a fact of history that the Roman Empire was preaching its own gospel decades before Christ began his ministry. They spoke of their own Lord (Caesar Augustus) who was in their mind the Son of God (Zeus) and brought a utopian age of peace, security and liberty to all mankind who had previously been living in fear of death and conflict. But the writings of the disciples and of Paul present Jesus Christ as the true and universal Lord and Son of God, who brings peace and reconciliation between God and humans, as well as between humans, and he sets up the church as the new alternative social order (not the Roman Empire). This new *euangelion* that Christ preached put him in direct opposition to the Roman Empire in general, and to Caesar Augustus in particular, who was emperor when Jesus was born. Could this be pointing in part to the violence that Christ was referring to in Matthew 11:12?

It is also a fact of history that the new kingdom of God—Christianity, heralded by the gospel—has indeed been growing to fantastic proportions (Matt 13:31, 33), and has indeed been producing all kinds of "treasures, new and old" (Matt 13:47, 52): not just spiritual benefits (forgiveness; salvation; healing), but also societal ones (hospitals; schools; humanitarian efforts; movements to abolish slavery; peace-and-reconciliation commissions after a genocide; the laws which govern many countries). It has been casting nets to catch "fish" (Matt 13:47; see Matt 4:19), and has been preparing a banquet (Matt 22:2). This is the ongoing fulfillment of Christ's prayer: "Thy kingdom come, thy will be done, on earth as it is in heaven" (Matt 6:10). And John the Revelator tells us the outcome of the competition between these two kingdoms: "The kingdom of the world has become the kingdom

---

117. Kim, *Christ and Caesar*.

of our Lord and of his Messiah, and he will reign forever and ever" (Rev 11:15). Prominent theologians such as N. T. Wright now call for such a reinterpretation of our understanding of what and where "heaven" is, challenging in particular the longstanding notion that our Christian faith is all about escaping earth in order to make it to some far-off place called heaven: instead, they suggest those passages teach about Christ's mission being to establish the kingdom of heaven on earth.[118]

So to bring this section full circle: did Christ in fact teach us anything about some non-physical, spiritual place far removed from earth, where God and the angels dwell? I do not think that question is so easy and straightforward to answer.

The Intermediate State and Purgatory

Several Christian traditions have developed distinct theologies pertaining to an existence between that on earth and that in heaven/hell. These are the "Intermediate State" and "Purgatory."

The "Intermediate State"—also referred to as "Soul Sleep" and "Abraham's bosom" (see older translations of Luke 16:22)—is a theological concept pertaining to that human existence which bridges the putting off of the physical body (biological death) and the putting on of the spiritual body (final resurrection). It is not held by all Christian traditions. Moreland and Rae argue vigorously for this concept, and describe three different views about the existence of the soul immediately following the death of the mortal body: (*i*) the "temporary disembodiment" position (to which Moreland and Rae themselves ascribe), in which the immortal, immaterial, and yet still conscious soul continues to exist in an unnatural and incomplete disembodied state while awaiting a new and final resurrection body; (*ii*) the "extinction-recreation" position, in which the person ceases to exist (and is therefore not conscious), but is later reconstituted at the Parousia when the new resurrection body is created (this position would be consistent with many monist views, particularly those in which the soul is an emergent property of the brain); (*iii*) the "immediate resurrection" position, in which the individual is immediately given their new resurrection body.[119]

Scriptural support for the temporary disembodiment position includes the passages referred to above for the ancient Hebrew belief of the dead—the *rephaim*, or the "shades"—dwelling "in Sheol in a kind of lethargic mode of existence marked by continuity of personal identity and the

---

118. Wright, *Surprised by Hope*; Wright, *Day the Revolution Began*.
119. Moreland and Rae, *Body and Soul*, 26–27.

capacity for being awakened and engaging in interpersonal discourse."[120] Jesus refers to the ancient Hebrew patriarchs as being still alive (Matt 22: 32; Mark 12:26–27; Luke 20:38). Proponents also point to other passages which refer to events that take place before the Second Coming of Christ (the "Parousia") which ushers in the Great Resurrection, and the Final Judgment. For example, Jesus' parable of the rich man and Lazarus (Luke 16:19–31) features two men in the "bosom of Abraham," an afterlife existence which precedes the Parousia (since the brothers of the rich man are still alive on earth). However, it is hard to call this a disembodied state, given that these two still have body parts (one a finger, and the other a tongue) and can still experience sensations which one normally associates with the body (the burning of a fire, and the cooling of water). Proponents also point to Jesus telling the thief on the cross about the paradise which awaits them later that day (Luke 23:40–43; which clearly preceded the Parousia). Paul provides numerous references to what some see as the Intermediate State: he hopes to live until the second coming of Christ in order to be immediately clothed with his heavenly body and thus avoid the "nakedness" of putting off the mortal body and awaiting the Parousia (2 Cor 5:1–10; also see Phil 1:21–24);[121] believers who have died are referred to as being "asleep," awaiting the Parousia (1 Cor 15:18, 20, 51; 1 Thess 4:15, 5:10). Referring to people who had died as being merely asleep—the Intermediate State?—was also done by Daniel (Dan 12:2), Job (Job 3:11–17), Jesus (John 11:11–14), and Luke (Acts 7:60). Peter mentions Jesus preaching to imprisoned spirits immediately following the death of his physical body (1 Peter 3:18–19).[122] John the Revelator describes the disembodied souls of the martyrs awaiting the Final Judgment (Rev 6:9–11).

An entirely different theology arose pertaining to a separate *place* which is intermediate between earth and heaven/hell: Purgatory. The rationale behind positing such a place is fairly straightforward. First, heaven is a place of perfection and the complete absence of sin and impurity: John the Revelator writes about the New Jerusalem that "nothing impure will ever enter it" (Rev 21:27), and the author of Hebrews writes, "Make every effort . . . to be holy; without holiness no one will see the Lord" (Heb 12:14). Second, everyone dies, but no one is perfect or sinless on their own merit when death claims them, even if they have indeed "made every effort" to pursue

---

120. Moreland and Rae, *Body and Soul*, 32; Cooper, *Body, Soul*, xv; Green, *Body, Soul*, 153.

121. Moreland and Rae, *Body and Soul*, 38.

122. Moreland and Rae, *Body and Soul*, 34.

holiness. As such, no one would ever "make it to heaven"[123] unless either: (*i*) God provides an interim place in which the individual can complete the process of sanctification until they are finally perfect and free of sin and/or have been thoroughly punished (the name "Purgatory" comes from the word "purge," referring to the purging or cleansing from sin, imperfections and impurities); or (*ii*) God removes their sin and gives them *his* holiness and perfection when they die.

The first of these two options is held by Roman Catholic and Eastern Orthodox Christians (as well as a minority of Protestants). Some posit that the reference to Sheol having compartments (Prov 7:27) alluded to a postmortem judgment in which "Sheol was the place where the souls of the wicked are judged and punished, and thereby purified, before they can be admitted to paradise."[124] This may be what the New Testament writers had in mind when they wrote that Christ descended into hell to preach to the imprisoned spirits (Eph 4:8; 1 Pet 3:19–20) and also visited paradise (Luke 23:43) before his ascent into heaven:[125] his visit to this "Purgatory" could be seen as an opportunity to encourage, convince, and provide direction to the residents of Purgatory, and possibly to escort those who had attained perfection into paradise/heaven itself.

It is further believed that the faithful living on earth can also play a role in that purification process for those waiting in Purgatory. It was at the time of the writing of the book of Daniel, in which the concept of bodily resurrection began to be more formally materialized within Jewish thinking, that we see believers praying for the dead (2 Macc 12:42–44; 1 Cor 15:29).[126] Later, the Roman Catholic Church allowed the selling of "Indulgences," which could reduce or even eliminate the amount of time which a deceased loved one had to spend in Purgatory. Even today, people will recite ritual prayers and light candles for deceased loved ones.

The Reformers (and now most Protestants) rejected this concept of Purgatory and embraced the second of the two options referred to above, replacing that concept with one in which Christ himself accepts all of God's wrath against the individual, receives their punishment, and pays all of their spiritual debt: "Penal Substitution."[127]

---

123. As noted above, theologians such as N. T. Wright are claiming that Christians should be emphasizing less how to "make it to heaven," and work harder on bringing heaven to earth: see Wright, *Day the Revolution*.

124. Pinker, "Sheol," 175; Green, *Body, Soul*, 160.

125. Harris, "Meaning of the Word Sheol," 129.

126. Beckwith, "Canon," 33.

127. Wright, *Day the Revolution*, 28–37.

The Resurrection Body

The concept of a resurrection body raises many questions. As already explained above, the Old Testament does not seem to address resurrection theology until we get to the book of Daniel (written during the exile), which describes only certain resurrected believers as shining like the stars (note that this resurrection is limited, not universal). There is no universal promise in Scripture of a restoration of the physicality of being (although noncanonical Hebrew texts like 2 *Maccabees* do clearly say that amputated body parts will be restored in the afterlife, and the dead bodies of martyrs will be raised to life) until we arrive at the gospel accounts of the resurrection of Christ, which speak to a very special case: the Messiah, whose whole mission was to be killed and then resurrected, thereby conquering death. These accounts paint a confusing picture of Christ's resurrected body: some interpret John's report to say that Jesus' body could pass through doors/walls (John 20:19, 26)[128] and yet his feet, hands, and side were apparently sufficiently physical that they could be touched and felt by his followers (Matt 28:9; Luke 24:39; John 20:17, 19), his feet seemed to meet enough resistance for him to be able to walk and stand, the clothes he was wearing found sufficient support to remain draped on his shoulders, and his vocal cords met enough resistance with air that he could speak; the resurrected Christ could sit, stand, and walk with his disciples as if still bound by gravity (Matt 28:9; Luke 24:15, 30, 36; John 20:14, 26) and yet float into the heavens as if not (Luke 24:51; Acts 1:9); he could speak with certain followers and not be recognized until a long period of time had passed (Luke 24:15–31; John 20:14–18); Jesus ate fish (Luke 24:41–43) and possibly Passover bread (Luke 24:29–30), but would not drink the fruit of the vine until his disciples were reunited with him in the heavenly kingdom (Matt 26:29). Jesus taught that others would also be resurrected, and would "be like the angels in heaven" (Matt 22:30; Mark 12:25; Luke 20:35).

For a short while it seems that relatively little thought was given to the question of physical resurrection of believers until the early church realized that more and more believers were dying even though Christ had not yet returned. Until that point, they had thought that his return was imminent—possibly days or weeks after his resurrection—to bring in the new Jewish kingdom. The fact that believers were already dying before the realization

---

128. The gospel writer does not actually say that Jesus passed through the wall; in both instances, the writer merely says that "though the doors were locked, Jesus came and stood among them." Other interpretations are equally possible, including Jesus simply somehow unlocking the main doors, or the room in which they were meeting having more than one entrance, or Jesus having slipped in with them unrecognized until later in the story (much like the two disciples on the road to Emmaus did not recognize him till hours after walking and talking intensely with him).

of that return raised serious questions in their minds in those early days of the church. This situation may have also brought into stark relief any latent divisions within their ranks owing to their Jewish cultural upbringing: at that point in church history, the Christian church comprised almost exclusively of Jews, all of whom were raised in a thoroughly Jewish home, some with the teachings of the Sadducees (that there is no resurrection; see 1 Cor 15:12), others with teachings of the Pharisees (there is indeed a resurrection). This motivated Paul to write his Letter to the Thessalonians—believed by many scholars to be the first of the books he wrote—to address this important question (1 Thess 4:13—5:11). Apparently the question needed to be revisited a few years later when he wrote his First Letter to the Corinthians (1 Cor 15:12-58). Other of Paul's writings then explore this question in increasing detail, but nonetheless leave a lot to the imagination and to speculation (which has since led to division within Christian ranks).

Drawing from the analogy of the green and leafy plant that arises out of the hard "dead" seed—and using that analogy to emphasize that the beginning and ending states are completely different from each other—Paul writes that "the body that is sown is perishable, it is raised imperishable; it is sown in dishonor, it is raised in glory; it is sown in weakness, it is raised in power; it is sown a natural body, it is raised a spiritual body. If there is a natural body, there is also a spiritual body" (1 Cor 15:42-44). His claim that "there is neither male nor female" when we are "clothed with Christ" (Gal 3:27-28) can be taken by some as literal and future-tense—there will be no gender in heaven (recall that Christ had already taught there would be no marriage in heaven [Matt 22:30; Mark 12:25; Luke 20:35])—but by others as metaphorical and eternal—on earth and in heaven, we do not discriminate on the basis of gender. Paul never clarifies what that "spiritual body" will look like. Peter describes the crucified Christ Jesus being "made alive in the Spirit" and preaching to imprisoned spirits, independently of his dead body (1 Pet 3:18-19).

The Christian emphasis on the resurrection (both of Christ and of believers) attests to the importance of the physical body. But what precisely is meant by "the body" which will be raised to new life? As noted above, a late second-century version of the Apostles' Creed specifically stated that *"the flesh"* would be resurrected.[129] This was later changed to "the dead," but even this change could still be interpreted to mean that it is the physical and biological body which is raised. However, that assertion raises numerous problems.

---

129. Endsjø, *Greek Resurrection*, 3.

First, clearly, the body is a collection of molecules arranged in a particular configuration which is unique to each individual. We have already seen how each of our bodies is very likely different in many ways from every other human body which has ever existed, and yet this truly unique body has been constantly changing throughout our lifetime (pages 115–116). Which one of these many different forms of our body will be resurrected? And what do we do with the Apostle Paul's assertion that, when we are clothed with Christ, there is no male and female (Gal 3:27–28), given that the many different versions of the body that we claimed as our own while living on earth very much had distinct degrees of male-ness or female-ness?

Second, our body is not simply a collection of *human* cells arranged in a particular configuration: we are hosts to a whole cosmopolitan community of bacteria, many of which are essential to our health and well-being. Not just a few bacteria: it has been estimated that there are ten times as many bacteria in and on our body than the number of our own human cells.[130] And not just of one kind of bacterium, but hundreds of different species of bacteria. Each one of us is actually a community of organisms—a bacterial-human chimera—one which is relatively unique to each of us. For example, only certain people might suffer their whole lives from halitosis (bad breath) or from bacterial ulcers, both of which are caused by unique types of bacteria which return to an individual even after treatment with antibiotics, and are not transmitted to other individuals whom they kiss. Some of these are just hitchhikers on our skin, in our lungs and elsewhere, and we can live quite well without them (in fact, our immune systems make significant effort in attacking them). But some are residents of our gastrointestinal tract and confer many benefits.[131]

> More than a billion years of mammalian-microbial coevolution has led to interdependency. As a result, the intestinal microbiota play a critical role in the maturation and continued education of the host immune response; provide protection against pathogen overgrowth; influence host-cell proliferation and vascularization; regulate intestinal endocrine functions, neurologic signaling, and bone density; provide a source of energy biogenesis (5 to 10% of daily host energy requirements); biosynthesize vitamins, neurotransmitters, and multiple other compounds with as yet unknown targets; metabolize bile salts.[132]

---

130. Reid and Greene, "FAQ: Human Microbiome."

131. Lynch and Pedersen, "Human Intestinal Microbiome," 2369–79; Conly and Stein, "Reduction of Vitamin K2," 531–39.

132. Lynch and Pedersen, "Human Intestinal Microbiome," 2370.

The study of the microbiota which inhabit our bodies, and their impact on our health and disease, is still a very new branch of science. We are learning, though, that we are also inhabited by fungi, viruses, and protists (the latter are similar to bacteria, but from an entirely different family of organisms): time will tell whether we find that they too play important roles in our complete well-being. It has been shown that proviral sequences which infected our distant ancestors—inserting themselves into the very DNA of those ancestors and thereby being handed down to subsequent generations—contributed to human brain development and accelerated our branching away from the other hominid lines.

Getting to the central point of this question, then: what will be "the body" which is resurrected? The early church fathers announced that it would be composed of the exact same molecules which we "owned" during our lifetime on earth (although they did not specify which of the many bodies we "owned" during the course of our earthly life). For example, Athenagoras (133–190 CE), a second-century Greek church father, wrote: "God will still see to it at the resurrection that the various particles of the flesh will be 'united again with one another,' so that 'they occupy the same place for the exact construction and formation of the same body.'"[133] Gregory of Nyssa (335–394 CE) explicitly said that in our resurrection, every "identical individual particle" which composes our bodies at death must return to us, else our resurrection will not be resurrection but "the creation of a new man": he insisted "the same man is to return to himself," down to "every single atom of his elements."[134] However, this is not physically possible even if they *had* specified which particular body from our lifetime they were referring to (for example, the one we were born with, or the one we died with). This is because the actual specific molecules of one's body have been swapped out hundreds (thousands?) of times over one's life history, and many (all?) of those particles have also been "owned" by other individuals in the past. Scott and Phinney refer to this as "the Cannibal Problem" (the particular problem of missionaries having been eaten by the cannibal tribe which later came to faith through their missionary efforts),[135] while Peters and Murphy refer to it as the problem of "chain consumption."[136] Peters also points out that, once again, Augustine thought about this problem before any of us and came up with what he thought was a solution:

---

133. Endsjø, *Greek Resurrection*, 2.

134. Gregory of Nyssa, "On the Soul and the Resurrection."

135. Scott and Phinney, "Relating Body and Soul," 96.

136. Peters, "Resurrection," 316–17; Murphy, "Nonreductive Physicalism: Philosophical Challenges," 108.

> For all the flesh which hunger has consumed finds its way into the air by evaporation, whence . . . God Almighty can recall it. That flesh, therefore, shall be restored to the [person] in whom it first became human flesh. For it must be looked upon as borrowed by the other person, and, like a pecuniary loan, must be returned to the lender.[137]

It is not clear how this problem is resolved for those individuals who "borrowed" atoms from previous owners: are those individuals resurrected with partial bodies, or with prosthetic substitutions? Instead, might holistic-resurrectionists refer only to some kind of coded template which can be imposed upon any other similar collection of molecules to form that unique body, thereby hearkening back to Plato's Forms and Ideas? Will that resurrected body include the several pounds of abdominal fat that one has been trying to lose for the past decade, or a receding hairline? Will it have all the accumulated injuries and defects—scars, limb amputations, or even the neurodegeneration seen in diseases such as Alzheimer's, schizophrenia, or even routine old age—"fixed" or reversed back to the original status of the pre-death body (and if so, to which version of the pre-death body)? Or will the resurrected body be an idealized body? If the latter is the case, will all resurrected bodies then look the same? Questions such as these may seem farcical or unnecessary to some readers: but they become increasingly important in proportion to the degree that one insists upon a physical/bodily resurrection.

Also, the claim that our resurrected body will constitute the very same molecules or coded template implies that it will look like and function like our current earthly body. That includes not only the bacteria and fungi in our body which are essential to our life on earth, but also a whole series of human cell types, organs, and mechanisms which again are necessary for life on earth but presumably not for the afterlife, and which nonetheless constitute a tremendous fraction of our body mass:

- our immune system and blood clotting system (altogether a tremendous fraction of our blood cells, plus several body organs) which become extraneous if there is no more disease, injury, or death;
- that part of our nervous system and peripheral sensors which responds to pain, hunger, thirst, stress, and other such "negative" sensations (again, a sizeable fraction of that system) which might never be experienced in heaven;

---

137. Augustine, *City of God*, XXII:20.

- the various cell types and mechanisms which are responsible for tissue regeneration and repair (fibroblasts, osteoclasts, osteoblasts; macrophages, nerve cells, and many others);
- the various nervous, glandular, and cellular systems which set up our body clocks and thereby orchestrate various body functions, since it is said there will be no time in heaven;
- there would be a major reconfiguration and down-sizing of our digestive tract, since much of that is needed to extract building materials and energy to contribute to tissue regeneration and repair, and in part doing that by digesting animal meat (some might say there will be no more carnivorous activity in heaven . . . would that mean that even certain teeth become unwarranted?);
- the various organs and systems which constitute our gender and serve only in sexual reproduction.

None of these cell types and organs listed here will be needed in an afterlife where there is no death, disease, degeneration, injury, or male/female genders.

Or will the resurrected body in fact be immaterial—Paul writes about it being a "spiritual body" in contrast to a "natural body" or "physical body" (1 Cor 15:44)—which then impacts interpretations of scriptural passages such as the future banquet that Jesus anticipates (Matt 26:29), or the "place" he promised to prepare (John 14:2–3) since the immaterial does not need a "place" to inhabit, or his prohibition against Mary touching his (immaterial) feet (John 20:17), or the immaterial New Jerusalem which John described in precise materialistic architectural detail (Rev 21:15–18). The Apostle Paul proposes that we will be given an entirely new kind of body; however, we have very little description from his writings or any other scriptural texts regarding the nature of that new kind of body. Many scholars agree that when Paul contrasts "physical body" and "spiritual body," he is not focusing specifically on the fact that the former is material and the latter immaterial, but rather that the former *is animated* by the soul while the latter *is animated by* the spirit.[138] James H. Charlesworth suggests (I would hope he did so tongue-in-cheek fashion, given that he is a New Testament language scholar, and not a nuclear physicist) that Einstein's famous equation which directly relates matter with energy—$E=mc^2$—might possibly shed light on the relationship between our physical body and spiritual body.[139] N. T. Wright writes: "God

---

138. Wright, "Transforming Reality," 120; Borg, "Truth of Easter," 133.
139. Charlesworth, "Resurrection," 171.

will download our software onto his hardware until the time when he gives us new hardware to run the software again."[140]

Summary

Compared to the New Testament, the Old Testament provides very little detail regarding the eternal destiny of the dead, the afterlife, or the nature of our immaterial being. The earliest texts do not seem to describe any kind of an afterlife judgment scenario, nor indeed any kind of distinct outcomes for the wicked and the righteous. Instead, *everyone* goes to Sheol—a gloomy place of nonexistence and complete abandonment—even though the garden of Eden or paradise have already been introduced in the opening chapters of Genesis as more than suitable places for an eternal postmortem existence. Even God himself at the very beginning of biblical history only warned humans that the outcome of reaching for the forbidden fruit would be death (Gen 2:17),[141] and in sentencing them for breaking that commandment he only said "for dust you are, and to dust you will return" (Gen 3:19; this is reiterated in Pss 90:3; 103:14; Eccl 3:20): there is no mention of a hell, or flames, or eternal punishment. Many of these views of the afterlife are similar to (or even echo) those of the Akkadians, Sumerians, Egyptians, and ancient Greeks.

The fact that so many biblical speakers apparently returned from the "Land of No Return," or were able to continue to do things (cry out to God, or even to shout and sing [Isa 26:19]), have feelings (despair; anguish; gloom; hope for deliverance) and experience their surroundings (darkness; maggots) while in the "Land of the Dead" suggests that Sheol might have been more a metaphor than an actual place. Sheol may also refer to being cut off from YHWH and from ancestry/descendants, adrift and lost in some kind of nonexistence. Vague hints of an afterlife can be glimpsed in later passages of the Old Testament.

Much later yet, the first clear description of an afterlife existence is found in the twelfth chapter of Daniel, where *some of* the righteous are rewarded (an eternal life with YHWH, in a seemingly celestial place) and *some of* the wicked punished (a life of shame and everlasting contempt). There is good reason to think that these ideas were again borrowed from the surrounding culture, particularly the Zoroastrians of the Persian Empire.

---

140. Wright, *Surprised by Hope*, 163.

141. Again, this is worded as an outcome or consequence, rather than as a punishment; see page 43.

This marks the beginning of a period of intense interest in and speculation upon the concepts of resurrection, postmortem judgment, and an afterlife by Hebrew thinkers. At the same time, the Greek empire brought its own globe-dominating influence upon the discussion, and all these Jewish ideas become Hellenized and later reinterpreted by Christian thinkers. Ultimately, the Hebrew *gn-'dn* and *Sheol* evolved into the Greek *paradeisos* and *Hades*, and those in turn into the Christian heaven and hell. Later thinkers then introduced the novel concepts of the "Intermediate State" and "Purgatory."

CHAPTER 5

Change

Two main themes emerge from this synthesis. First, it is undeniable that Judeo-Christian thinking on several key theological/philosophical points—human ontology, death, resurrection, afterlife, soul, etc.—has evolved considerably over the past several millennia. Second, that that evolution of our thinking was heavily influenced by our surrounding cultural neighbors, including the Babylonians, Egyptians, Zoroastrians, Greeks, and now more recently, the scientific enterprise and modern philosophy.

How Has Judeo-Christian Thinking Evolved?

Again, our theology has evolved: historical and literary scholarship unequivocally documents it. At first glance, it would seem that this evolution was not merely embellishment on a bare framework, nor a maturation of an undeveloped story. Instead, it has taken a number of sharp turns in new directions—in some cases one hundred and eighty degrees—and sometimes involved discarding whole portions of that which had been built up.

For several thousand years after leaving Africa, humans in the entire Mesopotamian region developed a theology which may have varied in particular details from one region or people group to the next, but from a broad perspective was relatively homogenous. For example, focusing on the subject of this book, the Akkadians, Sumerian, Egyptians, Greeks, Persians, and many others all believed that humans were created by a collaboration of gods, who mixed dirt with some divine element, but not for the purpose of any intimate relationship with humans. Upon death, humans were relegated to some kind of dreary underground existence, cut off from the gods and from any kind of community. Over time, those civilizations began to develop their mythology and their understanding of anthropogony and human ontology, with respect to both the present life and the afterlife. Abram's family and line of descendants lived in this ancient Near Eastern context, but discovered or developed their own quite unique theology.

But this ancient Hebrew theology/philosophy—Anthropogony 1.0—took a dramatic turn when it encountered and interacted with classical Greek philosophy. Anthropogony 2.0 was quite different in many respects:

Monism versus Dualism

Humans morphed from being an inanimate lump of clay to a *nephesh*—a holistic unity of life, body and consciousness—but then into a duality (body and soul) or even a trichotomy (body, soul, and spirit), and that in turn became ever more complicated with ideas of a multiplicity of souls (various iterations of humans having three or more souls) and the indwelling of various immaterial beings. In a large compendium of essays focusing on human ontology, Murphy stated that "the Hebrew idea of personality is that of an animated body, in contrast to the Greek [Christian] idea of an incarnated soul."[1] To be fair, though, her coauthor gave three reasons why it would be mistaken to contrast a "Hebraic view" of the human person with a "Greek view": the matter is not quite that black and white.[2]

The *Imago Dei*

The understanding of the *imago Dei* alluded to in Genesis changed completely: it had been a concept which was corporate in expression and rooted in community and relationships, to one that was individualized (often solely for males) and pertained to specific cognitive abilities.

The Afterlife Existence

The ancient Hebrew expectation was of a dark, dusty, gloomy nonexistence in isolation from all loved ones including God himself, but this changed into a judgment scenario followed by eternal residence in either a place of blissful reward or one of punishment, fire, and torture. Theologians later added the concepts of an Intermediate State and of Purgatory to bridge a perceived gap between life on earth and life in eternity.

---

1. Murphy, introduction to *Neuroscience and the Person*, v.
2. Green, "Restoring the Human Person," 11; also see Green, *Body, Soul*, 52–53.

The Afterlife Setting

The garden of Eden became paradise and then heaven, while Sheol became the metaphorical Hebrew Gehenna, then the literal Greek Hades, and then to us "hell."

Immaterial Beings

Another related concept which evolved considerably, but one which we did not consider in detail in this synthesis, is that of other immaterial beings: *Elohim* (a plural term) became singularized (YHWH; the Lord; God), the nameless enemy (*HaSatan*, or "the accuser") became personified (Satan), and later we find a pantheon of spiritual beings (angels and demons) with various ranks and hierarchies. The nature of the Godhead changed radically: the Shema which a fully observant Hebrew heard recited every day—"Hear O Israel, the Lord is our God, the Lord is One"—reminded them of the oneness of God, while the Hellenized Christians spoke of Jesus also being God, and eventually they developed the concept of the Trinity.

The Goodness of Creation and of the Body

The book of Genesis emphasized the goodness of all created things and implied Adam and Eve being quite comfortable in their nakedness and enjoying the divine command to make children ("go out and have sex"), but the classical Greeks struggled with the idea that matter was evil and sensuality must be suppressed. Much of this transferred into the thinking of Hellenized Hebrews and Christians, who learned to war against their bodies and sought to be released from the latter in order to "be at home with the Lord" (2 Cor 5:8). This relatively soft form of dualism grew into various harder forms (asceticism; monasticism; Gnosticism; Manichaeism) together with practices which denied the body (forbidding jewelry, attractive clothing, or elaborate hairstyles; covering all parts of the body with full length, head-to-toe unicolor smocks/cloaks, leaving only the hands and faces exposed) and even self-punitive practices (wearing abrasive clothing; pilgrimages on hands and knees to "holy places"; giving up things for Lent; self-flagellation and beating of the breast).

The Fall from Grace

The primary event which separated God and his people was for millennia seen to be the golden calf incident at Sinai in which Israel rejected God and chose a manufactured substitute to worship instead. But the Hellenized Jews and Christians replaced this with the incident in the garden of Eden in which Adam and Eve made a choice to break a direct command of God in an attempt to be like God. In the process, those Hellenized believers introduced the concept of original sin which has so heavily influenced Christian thinking right to the present day. In the various iterations of this narrative, the central characters who were ejected into a lower earthly existence for displeasing God include: Adam and Eve (Gen 3); the king of Tyre (Ezek 28); the king of Assyria (Ezek 31); the Morningstar, Son of the Dawn (Isa 14); and Lucifer (Luke 10).

Our Ultimate Destination Goal

N. T. Wright argues that it was Greek thinking that shifted Judeo-Christian thinking away from restoration of Eden (heaven) on earth and onto finding a way to flee earth and make one's way to heaven:

> That vision of a nonbodily ultimate "heaven" is a direct legacy of Plato and those like the philosopher and biographer Plutarch, a younger contemporary of St. Paul, who interpreted Plato for his own day. It is Plutarch, not the New Testament (despite what one sometimes hears!), who suggested that humans in the present life are "exiled" from their true "home" in "heaven." That vision of the future—an ultimate glory that has left behind the present world of space, time and matter—sets the context for what, as we shall see, is a basically paganized vision of how one might attain such a future.[3]

> We have *Platonized* our eschatology (substituting "souls going to heaven" for the promised new creation) and have therefore *moralized* our anthropology (substituting a qualifying examination of moral performance for the biblical notion of the human vocation), with the result that we have *paganized* our soteriology, our understanding of "salvation" (substituting the idea of "God killing Jesus to satisfy his wrath" for the genuinely biblical notions we are about to explore).[4]

---

3. Wright, *Day the Revolution*, 33–34.
4. Wright, *Day the Revolution*, 147 (italics in the original; also see p. 234). This

And now, Anthropogony 2.0 has encountered modern science and is interacting with it, and that is leading us (forcing us?) to yet entirely new directions in Christian thinking: Anthropogony 3.0. The concept that humans were fashioned from a lump of clay—held universally throughout the Mesopotamian region for millennia, and also embraced by most Jews and Christians up until a few hundred years ago (and continues to be embraced in many of their circles even now)—has been widely replaced with the concept of biological evolution. There is still considerable uncertainty, dispute, turmoil, and even acrimonious division over certain concepts, especially with respect to human evolution: many Christians can embrace biological evolution up to a point, insisting that humans alone must have been specially created. Christians are understandably resistant to the consensus of materialistic physicalists who conclude that there is no place for the supernatural in our thinking and that science has found no evidence for the soul or the afterlife so therefore those ideas should be discarded. But we are nonetheless increasingly recognizing and acknowledging that there are conceptual inconsistencies in our theology:

- It is one thing to dogmatically insist and maintain that humans comprise in part an immaterial component, but *how* does that immaterial spirit interact with our material body? In fact, *when* does that immaterial component first interact with the material body of an individual? Similarly, when did that immaterial component first begin to interact corporately with our species? How can it continue to exist when the body dies?

- It is one thing to maintain that humans can experience a postmortem resurrection, but exactly *what* is resurrected? Is it the very molecules that we "owned" before we died? If so, which ones, since all of them have been replaced numerous times during our lifetime, and have been "owned" by other believers? Which configuration of molecules will we receive or adopt: the body we were born with, or the one we had when we reached our peak (however one might define that), or the one we died with, or an idealized one?

- It is one thing to maintain that there is a "place" in which we can enjoy God's presence in eternity, but God is not bounded by space or time, nor is the immaterial component of our existence. So "where" would heaven or hell be found (the quotations are intended to emphasize the absurdity of asking about the three-dimensional co-ordinates of an

---

excerpt is an incredibly dense distillation of a very important reformulation of Christian theology, and I highly recommend that the reader consult the entire volume.

immaterial place)? Or should we instead be looking forward to God eventually establishing a form of heaven here on earth? (As Jesus himself put it: "Thy kingdom come . . . on earth as it is in heaven").[5]

Why Has Judeo-Christian Thinking Evolved?

As if it has not already been stated numerous times: it is clear now that our theology regarding human ontology and the afterlife has evolved. I would argue that this is the case not because God has been changing his mind about these matters. Instead, might this evolution reflect the fact that we have been in the process of discovering, revising, and/or creating these ideas?

Many ascribe to the idea that God chose to reveal truth in small increments because humans were not ready for the full revelation all at once. For example, modern-day critics of Christianity will point to the Levitical passages which prescribe laws that seem to accept and condone slavery, the subjugation of women, and the devaluation of children and foreigners, as well as historical passages which have God commanding the wholesale slaughter of people groups (including their women and children and sometimes even their animals). The Christian apologetic response to their criticisms and challenges has often been that the fledgling Israelite nation had just been liberated from bondage in Egypt, had been wandering around the desert for a while before moving into a conquered land in order to build their own society, and was not sociologically ready, intellectually sophisticated enough, nor logistically prepared for the concepts of full equality for all humans and more humane approaches to social care, criminal prosecution, and international diplomacy.

A good example of this would be their arrival at Mt. Sinai to receive God's law. The very first expanded set of legislation that the nation of Israel received—after the pithy Ten Commandments which most of Western society will recognize—focused squarely upon slavery (which makes sense, given that they had just come out of four hundred years of slavery themselves). And rather than being told simply to put an end to all forms of slavery, they are given laws to control it and limit it, using as case examples the particular situations of buying one's fellow Hebrew neighbor as slave, or selling one's daughter into slavery. The apologetic explanation for this today is something to the effect that: in their society and situation, very poor people might be forced to resort to extreme measures like selling themselves or their children into slavery in order to survive, and in the process put those slaves in extremely vulnerable positions of potential abuse and exploitation.

5. Wright, *Day the Revolution*, 147, 171.

The apologist will then emphasize that these laws had the best interests of the slaves in mind, and were a radical improvement over the norms for that time, since it enshrined limitations around those forms of slavery and protections for the slaves. Another set of laws which follow shortly thereafter put in place certain protections for a woman who might have been raped, including the stipulation that the rapist must then marry his victim with no option to divorce (the latter was permitted in certain other situations, but not this one). Again, it is emphasized that this protected the woman from being rejected by her community and being left defenseless against further assault or starvation, and therefore these laws once again represented "radical improvements over the norms for that time."

When the apologist is asked why a better solution would not be for the community to be instructed to come around those victims, show them unconditional love, and support them and provide for their needs, the response is inevitably that the people of that time were not ready for such a profound change. I am not fully convinced by or comfortable with that explanation. But for those who want scriptural evidence to defend a claim that God might work through partial solutions and reveal partial truths—otherwise referred to as Progressive Revelation—this would be one such precedent and exemplar.

It could also be argued that the fuller revelations were "given as they were needed," or "when the right time had come." For example, it was noted above how the concept of resurrection changed dramatically when the faithful faced a whole new level of persecution, even to the point of extreme bodily harm and martyrdom. Perhaps, it could be said, they particularly needed that theological encouragement at that time in order to have hope and remain faithful. Likewise, the Apostle Paul writes of God revealing the Messiah "when the right time had fully come" (Rom 5:6; Gal 4:4).

Third, one could argue that God has been revealing himself and aspects of his truth to *all* human civilizations and people groups—the General Revelation of which Paul wrote in Romans 1:20, or the *logos spermatikos* of which Justin Martyr wrote, or Clement of Alexandria's concept of Christianity as a "third race" (see page 23)—and that he sometimes placed pagan mythologies and philosophies in front of the Jewish authors to serve as a lens by which they might see new truths, or truth from a new perspective. This too can be seen as a divine act of inspiration, or at least a divinely orchestrated revelation. Might this have been behind their emerging understanding of the afterlife, a change which alleviated the cognitive dissonance of the gloomy nonexistence lying together with the unrighteous in Sheol (the more Sumerian or Akkadian views) becoming a time/place of reward (the more Egyptian view) and even a bright, joyous and reembodied existence (the Zoroastrian view)?

These three explanations—Progressive Revelation, revelation given on a need-to-know basis, and the *logos spermatikos* or General Revelation—do not at all need to preclude Divine inspiration of Scripture and of Judeo-Christian theology. In fact, they all reflect a Divine-human collaboration which has been going on for millennia. In addition to (divine) revelation, it took (divine) inspiration to catalyze or motivate (human) introspection, (human) meditation, and the (human) experiences of life to incrementally develop Judeo-Christian thinking. The Bible was not a fully formed manual lowered down from heaven, but is as much a human document as it is a divine one in the same sense that many Christians view Jesus as being both human and yet divine. Of course, there is a downside to this dual-centered collaboration: it gave humans opportunity to distort the clear message. We have been involved in every stage of its inspiration, writing, translation, copying, and interpreting, and we've greatly increased the noise to signal ratio in so many ways. But God chose this route nonetheless, to make us part of the journey and to have ownership in the relationship. I have heard a wonderful analogy for this divine/human partnership: one of a parent preferring to bake a cake with the help of their preschool children, making a big mess of the kitchen and creating a lopsided cake, but enjoying and prioritizing the bonding, relationship-building and experience together.[6]

In addition to our insertions into Scripture, we have far more often inserted our own ideas into the developing edifice or cathedral of our theology and philosophy. Certainly this would be the case for specific minutiae such as the existence of seven or ten levels of heaven, the idea that females lack the *imago Dei* which males possess, or the concept of an "age of accountability" which gives infants and children a free pass into heaven. It may also have been the case for larger matters of conduct (the prohibition against drinking alcohol, or gambling?) or sacramental issues (circumcision? infant baptism?). Some might add to the list much more contentious issues which are currently rocking the church (the role of women in church leadership; homosexuality). These and other matters have been a major impetus behind the numerous schisms that have split the church.

Those who are less willing to accept the human role within the authorship of Scripture will naturally ask how are we to know where to draw the line between divine and human as we read/interpret our Bibles and/or develop our theology. That takes time, practice, discussion, prayer, and divine intervention ... and learning from mistakes. We see an excellent case study

---

6. Bruxy Cavey, personal communication.

of this in the birth and growth of the New Testament church. When Jesus' followers began to exercise their corporate independence after Jesus had been "taken up into heaven," they seemed to focus their attention upon their fellow *Jews*. And for good reason. Jesus himself was a *Jew*, and saw himself as "sent only to the lost sheep of Israel" (Matt 15:24). All of the twelve disciples whom he chose were *Jews*, and he sent them out with the command, "Do not go among the Gentiles or enter any town of the Samaritans; go rather to the lost sheep of Israel" to proclaim the good news (Matt 10:5–6). The religious texts that they read and studied were largely *Hebrew* ones, and contained numerous anecdotes of YHWH dealing almost exclusively with *Israel*, bringing a *Jewish* Messiah, and restoring *Israel* as the chosen people and premier nation. After Jesus left, all the disciples and most of the other followers were *all Jews* who had gathered in *Jerusalem* (the capital city of the *Jewish* nation). Peter preached the first Christian sermon to what is described as a largely *Jewish* audience (Acts 2:5, 14). "Every day they continued to meet together *in the temple courts*" (Acts 2:46; emphasis added), a very *Jewish* setting in which one can reasonably conclude that the target audience was primarily *Jewish*. For a while, it seemed that their compassionate efforts were concentrated primarily upon *Jewish* recipients, such that it became necessary to institute a board of deacons to oversee the distribution of care to the Greek widows and orphans (Acts 6:1–7).

It is true that Jesus had commanded them to "go and make disciples *of all nations*" (Matt 28:19; italics added for emphasis), but they seemed to not fully comprehend (or accept?) the open nature of the gospel message until the first Council of Jerusalem (Acts 15) where it was officially decided, on the basis of Peter's vision of the blanket lowered from heaven, that gentiles were indeed also welcome into the new faith. For a while though, it seemed that the gentile converts were expected to act like *Jews* (be circumcised; avoid eating meat offered to idols; practice the now reconfigured form of the Passover meal). Apostles such as Paul and Barnabas made long missionary journeys to distant cities far beyond the borders of Israel, although Paul's custom when he visited any new city was to go first to their local *synagogue* (Acts 17:2): old habits die hard. Eventually, however, the church did open up completely to people of all races and nationalities, and this in turn brought on a whole range of other theological changes with the introduction of Greek philosophies, as we have already documented above.

Lesser examples of changes in theology apparently brought on by more careful reconsideration include the church's strict prohibition against eating meat offered to idols (Acts 15:29) but later dropping this prohibition (1 Cor 8:4–13), or Paul's dogmatic statements about hair length for men and women which were to him so clearly evident within "the very nature of things" (1 Cor 11:14–15). Both are now dismissed as just cultural artifacts.

It is a matter of debate—and the focus of this book—where one draws the line between divine inspiration and human discovery (or human invention). Recognizing that Scripture and theology both evolve over time allows us to reevaluate some of the older biblical passages. Jesus himself said, "You have heard that it was said, 'Eye for eye, and tooth for tooth.' But I tell you . . ." (Matt 5:38–39; but also see 5:40–48). It is now often said that we must "interpret Scripture through the lens of Jesus," or "have a Jesus-hermeneutic." When we find relatively ancient passages which advocate the killing of enemy tribes or eye-for-an-eye retribution juxtaposed against other more recent ones in which Christ teaches us to love our enemy and to forgive seventy times seven times, we should be guided by the latter pair of passages. Likewise, our understanding of human ontology and the afterlife should be informed by biblical texts, but the context(s) of their writing should be taken into account. Hermeneutics—the art of interpreting Scripture—is always influenced by the presumptions and worldview brought to the texts by the one doing the interpreting (as well as *inserted into* them by the authors and interpreters who wrote them).

Our Response to These Changes

The earliest Christians read their Bible in Greek, and interpreted their readings and thinking in a *Hellenized* Hebrew culture, if not a Greco-Roman one. As such, when they read about Sheol and the underworld, it would be natural for them to have images of the Greek Hades in mind. When they read about the garden of Eden, they would envision paradise (a concept which comes to us from the Persians) or the Elysian Fields of the Greeks. How could they *not* do this?

Now we find ourselves looking back at the words written by the ancient Hebrews and the New Testament authors, but do so through lenses tinted by modern science and postmodernism. An ever-accelerating flood of scientific discoveries has been constantly coming up against Christian theology. In the sixteenth century, astronomy and mathematics challenged the "biblical" model of the cosmos which the church had built upon a literal reading of various scriptural passages (page 69–70). European mariners exploring distant continents described races of people whose very existence challenged descent from a primal couple (since they were not sufficiently Caucasian, as per any painting from the Middle Ages and early Renaissance period) as well as a recent global flood (their ancestors could not possibly have survived it, nor had enough time to travel to such distant and isolated lands and set up such elaborate and distinct societies).[7]

The pace of clashes between faith and science accelerated dramatically in the second half of the nineteenth century, which saw scientific discoveries challenging the historicity of Adam and Eve as the first hominids (the discovery of the first Neanderthals in 1856), the means by which life forms came to be (publication of Charles Darwin's *Origin of Species* in 1859), the complete separation between species (the discovery in 1861 of a fossilized *Archaeopteryx*, a transitional species linking reptiles and birds), the young age of the earth (the discovery of radioactivity in 1896, which gave us radiometric dating), the global flood (a variety of geological discoveries throughout that period and later), the authenticity of the first eleven chapters of Genesis (the archaeological discovery in 1870 of Babylonian artifacts and clay tablets, including the Epic of Gilgamesh),[8] and the introduction of a new interpretive approach to the Old Testament texts which completely undermined a simple, literal reading of the latter (Julius Wellhausen and others in 1883).

The twentieth and twenty-first centuries have been no less forceful, with discoveries describing and explaining the very large (big bang cosmology) to the very small (quantum mechanics), the highly deterministic (genetics) to the bewilderingly chaotic (chaos theory), and everything in between. These scientific advances have been at times incredibly disorienting for those who hold to Christian theology, and the church has responded in a variety of ways. What has been the response to this conflict between faith and science? More importantly, what *should be* that response?

Discredit the Science

Some resort to flat denial of scientific findings when the latter come up against Christian belief. For example, defiant dogmatism is apparent in the opening lines of the controversial textbook *Biology for Christian Schools*,[9] which states, "If [scientific] conclusions contradict the Word of God, the conclusions are wrong, no matter how many scientific facts may appear to back them." This denial is also evident in the very large proportion of Christians—in certain Christian circles, this amounts to an overwhelming majority—who insist that humans did *not* evolve from a line which includes chimpanzees even when 98 percent of scientists *do* accept the evidence that this is the case (see page 66). Or they may demonize the scientific

7. Livingstone, *Adam's Ancestors*.
8. Scobie, "History of Biblical," 15.
9. Pinkston, *Biology for Christian Schools*, 1.

enterprise, and sow mistrust of it, by saying, "Science is always changing." They will refer to scientific studies which come to one conclusion, and other studies which come to a different or even opposite conclusion (for example, coffee or wine being good for one's health). Or to theories which were in vogue at one time but have since fallen out of favor (for example, the "science" of phrenology; page 63). And they may then juxtapose that against their perception of the relative bedrock constancy of Christian theology by quoting from the book of Hebrews: "but Jesus Christ is the same yesterday and today and forever" (Heb 13:8). However, they not only fail to fully understand the context of that particular biblical passage, they also fail to recognize that they have merely committed a bait-and-switch error: the question is not whether God or Christ change, but whether our *theology* has been changing. Very clearly the latter *has* been changing, just as much as science has been.

Elevate Scripture over Science

Rather than discredit or demonize science, some will resort to giving Scripture a degree of authority which it never claimed. They invoke the theological principles of inerrancy and infallibility: these too are a human development of Christian theology. The Old Testament writers, New Testament church, and the early church fathers did not seem to be concerned about matters of inerrancy and infallibility, but those concepts became matters of huge relevance for Christians ever since the nineteenth century. In some circles the twin concepts of inerrancy and infallibility have become a "shibboleth," a simple and unforgiving tool to separate Christians into two different tribes and kill those who are unable to see (say) things the way the ones in power do (see Judg 12: 4–6).

Give Up Faith?

Unfortunately, too many decide that the only appropriate response to these scientific discoveries is to give up their faith. Numerous polls and studies have documented this. For example, in another study conducted by Barna Research Group, 1,296 young people who had been regular churchgoers during their teen years but had since left the church were asked the reason for their decision to leave. Three prominent reasons given included "Christianity is anti-science" (25%); "been turned off by the creation-versus-evolution debate" (23%); and "churches are out of step with the scientific world we live in" (29%).[10] Exploring further the factors that contributed to young people dropping out of church involvement, Barna found college/university can be adversarial to faith for many (but not all):[11]

> The problem arises from the inadequacy of preparing young Christians for life beyond youth group.... Only a small minority of young Christians has been taught to think about matters of faith, calling, and culture. Fewer than one out of five have any idea how the Bible ought to inform their scholastic and professional interests. And most lack adult mentors or meaningful friendships with older Christians who can guide them through the inevitable questions that arise during the course of their studies. In other words, the university setting does not usually cause the disconnect; it exposes the shallow-faith problem of many young disciples.[12]

Another important reason given for leaving was that the church was unfriendly to those who doubt: "They do not feel safe admitting that sometimes Christianity does not make sense."[13] This author hopes that this book inspires Christian leaders to provide better mentorship and training in this area of faith, and open environments in which questions and doubts can be examined. The goal for this book is not to dispel the notions of God, spirit, heaven, or any other such aspects of traditional Christian theology, but instead to find a more relevant, complete—and perhaps even accurate—understanding of the human-Divine relationship.

Embrace Change

By now, it should be undeniable that Christian theology has been changing, as has scientific understanding of matters of great relevance to theologians. Both changes are a natural consequence of humans diligently working to discover the unknown, and of students learning a new and complicated subject. It can equally be said that the change in thinking is a product of teachers omitting the nuances and details of a complicated subject, or even simplifying the subject they are teaching to the point of distorting it until the students are ready for the whole truth. One example of this would be sex education at the various levels of elementary school and then high

---

10. Barna, "Six Reasons," reason #3.
11. Barna, "Most Twentysomethings," para. 5.
12. Barna, "Five Myths," myth #3.
13. Barna, "Six Reasons," reason #6.

school. Another would be teaching about imaginary numbers in mathematics: elementary students learn that multiplying any number by itself will *never* produce a negative number (3 times 3 equals 9, but -3 times -3 also produces plus 9), but years later they will encounter higher forms of math used in many branches of science and technology, including the concept of imaginary numbers which completely break the rule just mentioned ($3i$ times $3i$ produces negative 9, not plus 9). Changes do not need to be feared. They do not need to threaten us. We do not need to question or challenge the doctrine of divine inspiration or revelation, nor react defensively against new questions or new discoveries. Again, we the church have responded inappropriately to such changes in the past, and have often had to reverse our dogmatic opposition and recant our vigorous denial. Three hundred years after the "Galileo Affair," the Roman Catholic Church offered an apology of sorts for the way it responded to scientific facts which challenged theological doctrine. I would reiterate Goheen and Bartholomew's conclusion that "it didn't have to be this way"[14] when it came to the structure of the cosmos (see page 69–70), nor does it need to be this way for the present issue of human ontology and the afterlife. As noted above (page 23), Clement of Alexandria was quite comfortable with Christian thinking being the fulfillment of Hebrew theology and Greek philosophy, the latter being the science of his day (and quite pagan at heart no less). John Calvin advocated interpreting Scripture through the latest science of his day, not only because this would lead to a more accurate interpretation of Scripture and more complete understanding of truth, but also because the science would stimulate wonder and worship. Echoing Clement of Alexandria, he said that the Holy Spirit is the author of all truth, and that "pagan" science is also truth and capable of revealing things about God (although he emphasized that one still needs theology and the Bible to make proper sense of it).[15]

Changes Going Forward: Ethical and Spiritual Considerations

A more theologically and scientifically informed understanding of human ontology will better equip us for a variety of ethical decisions which are increasingly facing us, ones which previous generations never had to contend with. A number of these are explored in detail by D. Gareth Jones,[16] and are briefly summarized here.

14. Goheen and Bartholomew, *Living at the Crossroads*, 89.
15. Calvin, *Institutes*, 2.2.15, 273–74; 2.2.15–16, 273–75.
16. Jones, "Emergence of Persons," 11–33.

Given how mind and consciousness are so closely tied to brain function, how does one's perception of the personhood of a loved one change as they progressively lose cognitive functions in neurodegenerative situations such as Alzheimer's disease, dementia, and even advanced old age? Or the victim of a car accident who is suddenly robbed of all brain activity and enters a persistent vegetative state but whose bodily functions can be maintained on life support for years at a time? Increasingly, families are having to make the excruciating decision of when to end life support for such a loved one. The importance of this question has been dramatically amplified by the discovery that people can endure a terrifying condition known as "locked-in syndrome," in which the patient is fully able to receive and process sensory input and to be cognitively aware of their environment, but are completely unable to make any kind of signal whatsoever to their caregivers that they are in fact awake and aware. They can on all accounts appear to be in a comatose state, but we have now begun to learn how to use electrical brain wave recordings to communicate with them.[17]

One can examine this same question in the reverse direction: if there is a neurologically defined end-point to human life, can there be a neurologically defined starting point for the same? That is, if there is such a thing as brain-death, can one also define brain-birth? This question is already at the heart of the debate over abortion: at what point in gestation can one call the fetus "human," when does it gain consciousness, and when is it capable of suffering? Is there a minimal amount of neuronal development and neural activity which can be defined as the threshold for human personhood? If so, does this need to be taken into consideration when we use pieces of brain tissue to treat neurodegenerative diseases like Parkinson's disease and Huntington's disease? This has already been done using fetal brain tissue, but theoretically could be done using adult brain tissue (to sidestep the ethical issues of infanticide), and has also been done using animal brain tissue. Should it be taken into consideration when scientists culture human brain cells for experimental research—generating vast amounts of brain cells which far exceed those of a normal intact human brain—and then begin to reconstitute various combinations of brain cells to investigate how they interact?

Will a more detailed scientific understanding of mind and consciousness change the way we interpret situations which were previously seen to be interactions between humans and spiritual beings? Users of hallucinogenic drugs describe experiences with intense spiritual value which they sometimes characterize as "seeing God," when all that may have happened is that a neural cascade has been triggered by stimulation of $5HT_{2A}$-receptors (see pages

---

17. Chaudhary et al., "Brain-Computer Interface," e1002593.

103–105). Likewise, intense emotions can be evoked during meditation, music, or certain spiritual practices, all of which can induce a sense of awe, wonder, peace, and blissfulness which is otherwise attributed to an encounter with God. As we gain a greater understanding of the mechanisms underlying sleep and dreaming—distinct brain states in which neural processing is radically altered—will we need to revise our understanding of certain claims of God speaking to people through dreams? Conversely, how will manifestations which in the past were attributed to demon possession or demon oppression[18] be reinterpreted as we learn more about personality disorders and psychological dysfunction? For example, night terror (known clinically as *pavor nocturnus*) is a terrifying experience which occurs during a specific point in the normal sleep cycle (arousal from delta sleep, otherwise known as stage 3 of non-rapid eye movement, or non-REM, sleep) and causes terror and dread. In sleep paralysis, the otherwise awake person feels unable to move or speak and may hear things (including voices) or see things (including beings which they perceive as a demon sitting on their chest or a menacing figure lurking in the shadows) which altogether induce intense fear.

There are yet other questions which could not possibly have been anticipated by any scholars prior to the modern scientific period, let alone the past few decades. Some are talking today about transhumanism: the idea that humans can merge with their technology. They envision a future in which it might be possible to re-create all the synaptic connections in one's brain in some kind of physical electronic device and/or within some kind of immaterial computer program. In fact, some have currently had their brains frozen in liquid nitrogen, or made arrangements for that to be done when they die, in the hopes that their brains can be thawed and processed through that new technology. Others have recognized that the freezing and thawing will definitely induce some degree of damage (since water swells when it freezes, there will inevitably be some amount of cellular bursting and structural connections snapping), and for that reason are considering having their brains chemically treated (the colloquial terms here are "pickled" or "fixed") to better preserve microscopic structures. In both cases, even if the process of freezing or chemical treatment themselves could be optimized such that they caused no cellular damage in and of themselves, there would inevitably be some damage which resulted from the process of dying itself: the varying periods of oxygen-deprivation (anoxia) and lack of blood perfusion (ischemia). Would one therefore need to (want to) make the decision to start the freezing or chemical treatment before the moment of death? Would this decision need to be made before

18. Davies, "Spiritual Awareness," 131–32.

the medical condition which was leading to death had itself progressed too far? If so, could this lead to people choosing a carefully orchestrated suicide while they were still reasonably young and healthy in order to achieve a perfect state of immortality?

Another set of questions which could not have possibly been anticipated by the church during the first few centuries pertain to cell cloning and stem cell technology. We have already encountered this above (see page 87). Today, it is possible to clone cells. The copies of those cells are generally in what scientists call a differentiated state, meaning they already have a cell "type": clones of liver cells are liver cells, and clones from neural tissues are neurons. Stem cells are a less differentiated form of cell that can be cloned (multiplied) and then induced to take on different cell type appearances. Previously, stem cells were only available from fetal tissues: the latter begin from one fertilized egg and clonally expand into a mass of tissue, full of partially differentiated cells, each of which continue to clonally expand and then are induced by internally generated chemical signals to differentiate into the final cell types which altogether make a human baby. Now, however, it is becoming possible to take fully differentiated cells, such as those from a scraping of the mouth, and convert those into stem cells (which can then be turned into any kind of cell type). This new technology is now being used to not only create a homogenous mass of one particular cell type, but even to grow whole organs composed of many different cell types. This field is moving along at an incredible and ever-increasing pace. Projecting along the arc of those scientific advances, it will soon be possible to make an entire organism from a cheek scraping (although one would think—indeed, hope—that the technology would be held back by discussions of ethics and morality). If/when that day comes, one could argue that those clones would have a mind and a personality, but would they have a soul? If yes, would each have a unique soul? Would those clones have full and independent human rights?

Strange and disturbing questions indeed.

Concluding Statement

Judeo-Christian thinking has been in flux for thousands of years, and many more changes are on the horizon. Although a large proportion of scholars may agree that dualism "does not work," the fact is that most of the people in the pews of our churches are dualists in worldview. Likewise, most nonbelievers are monists, and believe that dualism is a core doctrine of Christianity. This represents, then, a major dividing line between believers and nonbelievers, as well as between the laity and the academy. One reason, then, to look for scientific explanations of human ontology that do not compromise Christian faith is to find unity with fellow believers who place the "Book of God's Works" on par with the "Book of God's Word." Another reason is that this will remove a philosophical barrier between believers and others who are put off by the mystical and the spiritual. It is only entrenched tradition and human interpretation of Scripture which separates these groups, not Scripture itself. It is entirely possible to hold a Christian faith which is compatible with even the latest developments of science, as long as we remain open and think critically. In Joel B. Green's words: "I demonstrate that those views of the human person which are consistent with what we are learning from the natural sciences present no fundamental challenge to biblical faith."[19]

I hope that the reader can find in this book some models of human ontology which do not depend upon a supernatural component that is by definition untouchable by scientific investigation, and find ones which can get over the hurdle of Descartes's question of how the material and immaterial interact.

---

19. Green, "A Response," 194.

About the Author

Dr. Luke J. Janssen is a professor in the Department of Medicine of a leading Canadian university, and holds a bachelor of science degree in biochemistry (1984), a master's degree in medical science (1987), and a doctorate of philosophy in medical sciences (physiology and pharmacology, 1990). He has published 148 scientific research articles and book chapters and presented his research before many national and international scientific and pharmaceutical industrial audiences.

He has also completed a master's degree in theological studies (2019) and published two other books and a paper in the area of faith-science apologetics, and maintains a blog that addresses many of the same questions (https://lukejjanssen.wordpress.com). He has served for over thirty years in various church ministries: these have included service on a board of elders, leadership of youth groups and young adult groups, teaching in Sunday school classes (children, youth, young adult, and adult classes), and numerous ministries of service.

Bibliography

Abusch, Tzvi. "Ghost and God: Some Observations on a Babylonian Understanding of Human Nature." In *Self, Soul and Body in Religious Experience*, edited by A. I. Baumgarten et al., 363–83. Leiden: Brill, 1998.
Adler, Cyrus, et al., eds. *Jewish Encyclopedia*. Vol. 6. New York: KTAV, 1941.
Al-Khalili, Jim, and Johnjoe McFadden. *Life on the Edge: The Coming of Age of Quantum Biology*. London: Bantam, 2014.
Allen, Diogenes. "Persons in Philosophical and Biblical Perspective." In *From Cells to Souls—and Beyond: Changing Portraits of Human Nature*, edited by Malcolm Jeeves, 165–78. Grand Rapids: Eerdmans, 2004.
American Association for the Advancement of Science. "About AAAS." http://www.aaas.org/about-aaas.
Anderson, Michael R., et al. "Genetic Correlates of Spirituality/Religion and Depression: A Study in Offspring and Grandchildren at High and Low Familial Risk for Depression." *Spirituality in Clinical Practice* 4 (2017) 43–63.
Anderson, Ray S. "On Being Human: The Spiritual Saga of a Creaturely Soul." In *Whatever Happened to the Soul?*, edited by Warren S. Brown et al., 175–94. Minneapolis: Fortress, 1998.
Aquinas, Thomas. *Summa Contra Gentiles*. https://dhspriory.org/thomas/ContraGentiles.htm.
Ashwin-Siejkowski, Piotr. *Clement of Alexandria: A Project of Christian Perfection*. T. & T. Clark, 2008.
Athanasius. *On the Incarnation*. https://www.ccel.org/ccel/athanasius/incarnation.
Auffarth, Christoph. "Paradise Now—but for the Wall Between: Some Remarks on Paradise in the Middle Ages." In *The Creation of Man and Woman: Interpretations of the Biblical Narratives in Jewish and Christian traditions*, edited by Gerard P. Luttikhuizen, 168–79. Leiden: Brill, 2000.
Augustine. *City of God*. http://www.newadvent.org/fathers/1201.htm.
———. *Confessions*. Book XI. http://www.newadvent.org/fathers/110111.htm.
———. *De doctrina Christiana*. http://www.newadvent.org/fathers/12022.htm.
———. *Letters*. Vol. 4. https://muse.jhu.edu/book/20857.
Ayala, Francisco J. "Human Nature: One Evolutionist's View." In *Whatever Happened to the Soul?*, edited by Warren S. Brown et al., 31–48. Minneapolis: Fortress, 1998.
Baars, Bernard J. "Global Workspace Theory of Consciousness: Toward a Cognitive Neuroscience of Human Experience." *Progress in Brain Research* 150 (2005) 45–53.

Baars, Bernard J., and Stan Franklin. "An Architectural Model of Conscious and Unconscious Brain Functions: Global Workspace Theory and IDA." *Neural Networks* 20 (2007) 955–61.

Baars, Bernard J., and David B. Edelman. "Consciousness, Biology and Quantum Hypotheses." *Physics of Life Reviews* 9 (2012) 285–94.

Ballerini, M., et al. "Interaction Ruling Animal Collective Behavior Depends on Topological rather than Metric Distance: Evidence from a Field Study." *Proceedings of the National Academy of Sciences* 105 (2008) 1232–37.

Bar, Shaul. "Grave Matters: Sheol in the Hebrew Bible." *Jewish Bible Quarterly* 43 (2015) 145–53.

Barbour, Ian G. *Religion and Science: Historical and Contemporary Issues.* San Francisco: HarperCollins, 1997.

Barbour, Ian G. "Neuroscience, Artificial Intelligence, and Human Nature: Theological and Philosophical Reflections." In *Neuroscience and the Person: Scientific Perspectives on Divine Action*, edited by Robert John Russell et al., 249–80. Vatican City: Vatican Observatory Publications, 1999.

Barna Group. "Five Myths about Young Adult Church Dropouts." November 15, 2011. https://www.barna.com/research/five-myths-about-young-adult-church-dropouts/.

———. "Most Twentysomethings Put Christianity on the Shelf Following Spiritually Active Teen Years." September 11, 2006. https://www.barna.com/research/most-twentysomethings-put-christianity-on-the-shelf-following-spiritually-active-teen-years/.

———. "Six Reasons Young Christians Leave Church." September 27, 2011. https://www.barna.com/research/six-reasons-young-christians-leave-church/.

———. "What Is Barna?" https://www.barna.com/about/.

Bauckham, Richard. "Paradise in the *Biblical Antiquities* of Pseudo-Philo." In *Paradise in Antiquity: Jewish and Christian Views*, edited by Markus Bockmuehl and Guy G. Stroumsa, 43–56. New York: Cambridge University Press, 2010.

Beauregard, Mario, and Vincent Paquette. "EEG Activity in Carmelite Nuns During a Mystical Experience." *Neuroscience Letters* 444 (2008) 1–4.

———. "Neural Correlates of a Mystical Experience in Carmelite Nuns." *Neuroscience Letters* 405 (2006) 186–90.

Beckwith, R. T. "The Canon of Scripture." In *New Dictionary of Biblical Theology*, edited by T. Desmond Alexander and Brian S. Rosner, 27–34. Downers Grove: InterVarsity, 2000.

Benjamins, Hendrik S. "Keeping Marriage Out of Paradise: The Creation of Man and Woman in Patristic Literature." In *The Creation of Man and Woman: Interpretations of the Biblical Narratives in Jewish and Christian Traditions*, edited by Gerard P. Luttikhuizen, 92–106. Leiden: Brill, 2000.

———. "Paradisiacal Life: The Story of Paradise in the Early Church." In *Paradise Interpreted: Representations of Biblical Paradise in Judaism and Christianity*, edited by Gerard P. Luttikhuizen, 153–67. Leiden: Brill, 1999.

Berger, Klaus. *Identity and Experience in the New Testament.* Minneapolis: Fortress, 2003.

Berry, R. J. "Natural Evil: Genesis, Romans, and Modern Science." *Perspectives on Science and Christian Faith* 68 (2016) 87–98.

Bingham, Jeffrey D. *Irenaeus' Use of Matthew's Gospel in Adversus Haereses.* Leuven: Peeters, 1998.

BioLogos editorial team. "A Survey of Clergy and Their Views on Origins." May 8, 2013. https://biologos.org/blogs/archive/a-survey-of-clergy-and-their-views-on-origins.

Bjork, Russell C. "Artificial Intelligence and the Soul." *Perspectives on Science and Christian Faith* 60 (2008) 95–102.

Bloch-Smith, Elizabeth. *Judahite Burial Practices and Beliefs about the Dead.* Sheffield, UK: University of Sheffield Press, 1992.

Bockmuehl, Markus. "Locating Paradise." In *Paradise in Antiquity: Jewish and Christian Views*, edited by Markus Bockmuehl and Guy G. Stroumsa, 192–209. New York: Cambridge University Press, 2010.

Boers, Hendrikus. "The Meaning of Christ's Resurrection in Paul." In *Resurrection: The Origin and Future of a Biblical Doctrine*, edited by James H. Charlesworth, 104–37. New York, T. & T. Clark, 2006.

Borg, Marcus. "The Truth of Easter." In *The Meaning of Jesus: Two Visions*, edited by N. T. Wright and Marcus Borg, 129–42. New York: HarperCollins, 1999.

Branson, Robert D. "Science, the Bible, and Human Anatomy." *Perspectives on Science and Christian Faith* 68 (2016) 229–36.

Bremmer, Jan N. "Pandora or the Creation of a Greek Eve." In *The Creation of Man and Woman: Interpretations of the Biblical Narratives in Jewish and Christian Traditions*, edited by Gerard P. Luttikhuizen, 19–33. Leiden: Brill, 2000.

———. "Paradise: From Persia, via Greece, into the Septuagint." In *Paradise Interpreted: Representations of Biblical Paradise in Judaism and Christianity*, edited by Gerard P. Luttikhuizen, 1–20. Leiden: Brill, 1999.

Brookes, Jennifer C., et al. "Could Humans Recognize Odor by Phonon Assisted Tunneling?" *Physical Review Letters* 98 (2007) 038101.1–4.

Brown, Samantha, et al. "Identification of a New Hominin Bone from Denisova Cave, Siberia Using Collagen Fingerprinting and Mitochondrial DNA Analysis." *Scientific Reports* 6 (2016) 23559.

Brown, Warren S. "Cognitive Contributions to Soul." In *Whatever Happened to the Soul?*, edited by Warren S. Brown et al., 99–125. Minneapolis: Fortress, 1998.

———. "Neurobiological Embodiment of Spirituality and Soul." In *From Cells to Souls—and Beyond: Changing Portraits of Human Nature*, edited by Malcolm Jeeves, 58–76. Grand Rapids: Eerdmans, 2004.

———. "Reconciling Scientific and Biblical Portraits of Human Nature." Conclusion to *Whatever Happened to the Soul?*, edited by Warren S. Brown et al., 213–28. Minneapolis: Fortress, 1998.

Burkert, Walter. *Greek Religion.* Cambridge: Harvard University Press, 1985.

Burns, Jeffrey M., and Russell H. Swerdlow. "Right Orbitofrontal Tumor with Pedophilia Symptom and Constructional Apraxia Sign." *Archives of Neurology* 60 (2003) 437–40.

Byrd, Michael F. *The Gospel of the Lord: How the Early Church Wrote the Story of Jesus.* Grand Rapids: Eerdmans, 2014.

Cahill, Thomas. *The Gifts of the Jews: How a Tribe of Desert Nomads Changed the Way Everyone Thinks and Feels.* New York: Bantam, 1998.

Calvin, John. *The Institutes of the Christian Religion.* Book 2. Translated by Henry Beveridge. Grand Rapids: Christian Classics Ethereal Library. https://www.ccel.org/ccel/calvin/institutes.toc.html.

Camus, Albert. *Christian Metaphysics and Neoplatonism.* Columbia: University of Missouri Press, 2007.

Charlesworth, James H. "The Origin and Development of Resurrection Beliefs." Conclusion to *Resurrection: The Origin and Future of a Biblical Doctrine,* edited by James H. Charlesworth, 218–31. New York: T. & T. Clark, 2006.

———. "Resurrection: The Dead Sea Scrolls and the New Testament." In *Resurrection: The Origin and Future of a Biblical Doctrine,* edited by James H. Charlesworth, 138–86. New York: T. & T. Clark, 2006.

———. "Where Does the Concept of Resurrection Appear and How Do We Know That?" In *Resurrection: The Origin and Future of a Biblical Doctrine,* edited by James H. Charlesworth, 1–21. New York: T. & T. Clark, 2006.

Chaudhary, Ujwal, et al. "Brain-Computer Interface-Based Communication in the Completely Locked-In State." *Public Library of Science Biology* 15 (2017) e1002593.

Childs, Brevard S. *Biblical Theology of the Old and New Testaments: Theological Reflection on the Christian Bible.* Minneapolis: Fortress, 1992.

———. *Old Testament Theology in a Canonical Context.* Philadelphia: Fortress, 1985.

Clayton, Philip. "Neuroscience, the Person, and God: An Emergentist Account." In *Neuroscience and the Person: Scientific Perspectives on Divine Action,* edited by Robert John Russell et al., 180–214. Vatican City: Vatican Observatory Publications, 1999.

Clouser, Roy. "Reading Genesis." *Perspectives on Science and Christian Faith* 68 (2016) 237–61.

Collini, Elisabetta, et al. "Coherently Wired Light-Harvesting in Photosynthetic Marine Algae at Ambient Temperature." *Nature* 463 (2010) 6444–47.

Collins, John J. *A Commentary on the Book of Daniel.* Minneapolis: Fortress, 1993.

Conly, John, and Kathryn E. Stein. "Reduction of Vitamin K2 Concentrations in Human Liver Associated with the Use of Broad Spectrum Antimicrobials." *Clinical and Investigative Medicine* 17 (1994) 531–39.

Cooper, John W. *Body, Soul, and Life Everlasting: Biblical Anthropology and the Monism-Dualism Debate.* 2nd ed. Grand Rapids: Eerdmans, 2000.

Cortez, Marc. *Theological Anthropology: A Guide for the Perplexed.* New York: T. & T. Clark, 2010.

Crenshaw, James L. "Love Is Stronger than Death: Intimations of Life beyond the Grave." In *Resurrection: The Origin and Future of a Biblical Doctrine,* edited by James H. Charlesworth, 53–78, New York: T. & T. Clark, 2006.

Crick, Francis. *The Astonishing Hypothesis: The Scientific Search for the Soul.* New York: Scribner, 1994.

Cyril. *Catechesis.* Catechetical Lecture 19. http://www.newadvent.org/fathers/310119.htm.

Dalley, Stephanie. *Myths from Mesopotamia: Creation, Flood, Gilgamesh and Others.* Oxford: Oxford University Press, 1989.

Davies, Douglas J. *Death, Ritual and Belief.* 2nd ed. London: Continuum, 2002.

Davies, Gaius. "Spiritual Awareness, Personality, and Illness." In *From Cells to Souls—and Beyond: Changing Portraits of Human Nature,* edited by Malcolm Jeeves, 123–45. Grand Rapids: Eerdmans, 2004.

Damrosch, David. *The Buried Book: The Loss and Rediscovery of the Great Epic of Gilgamesh*. New York: Holt, 2007.
D'Costa, Gavin. *Resurrection Reconsidered*. Oxford: Oneworld, 1996.
De Jager Meezenbroek, Eltica, et al. "Measuring Spirituality as a Universal Human Experience: A Review of Spirituality Questionnaires." *Journal of Religion and Health* 51 (2012) 336–54.
Dennett, Daniel C. *Consciousness Explained*. Boston: Little, Brown, 1991.
Dever, William G. *Who Were the Early Israelites and Where Did They Come From?* Grand Rapids: Eerdmans, 2006.
Dillon, John. "Origen and Plotinus: The Platonic Influence on Early Christianity." In *The Relationship between Neoplatonism and Christianity*, edited by Thomas Finan and Vincent Twomey, 7–26. Dublin: Four Courts, 1992.
Doblin, Rick. "Pahnke's 'Good Friday Experiment': A Long-Term Follow-Up and Methodological Critique." *Journal of Transpersonal Psychology* 23 (1991) 1–28.
Dunn, Geoffrey D. "Roman and North African Christianity." In *The Routledge Companion to Early Christian Thought*, edited by D. Jeffrey Bingham, 154–71. London: Routledge, 2010.
Eaves, Lindon. "Genetic and Social Influences on Religion and Values." In *From Cells to Souls—and Beyond: Changing Portraits of Human Nature*, edited by Malcolm Jeeves, 102–22. Grand Rapids: Eerdmans, 2004.
Eccles, John C., ed. *Brain and Conscious Experience*. Berlin: Springer-Verlag, 1966.
Edwards, Mark J. "Early Christianity and Philosophy." In *The Routledge Companion to Early Christian Thought*, edited by D. Jeffrey Bingham, 38–50. London: Routledge, 2010.
Elledge, C. D. "Resurrection of the Dead: Exploring Our Earliest Evidence Today." In *Resurrection: The Origin and Future of a Biblical Doctrine*, edited by James H. Charlesworth, 22–52, New York: T. & T. Clark, 2006.
Elsevier. "Discovery of Quantum Vibrations in 'Microtubules' inside Brain Neurons Supports Controversial Theory of Consciousness." *ScienceDaily*, January 16, 2014. https://www.sciencedaily.com/releases/2014/01/140116085105.htm.
Endsjø, Dag Øistein. *Greek Resurrection Beliefs and the Success of Christianity*. New York: Palgrave MacMillan, 2009.
Engel, Gregory S., et al. "Evidence for Wavelike Energy Transfer through Quantum Coherence in Photosynthetic Systems." *Nature* 446 (2007) 782–86.
Ferguson, Everett, ed. *Encyclopedia of Early Christianity*. New York: Garland, 1990.
Finkelstein, Israel, and Neil Asher Silberman. *The Bible Unearthed: Archaeology's New Vision of Ancient Israel and the Origin of Its Sacred Texts*. New York: Simon and Schuster, 2002.
Finlay, Graeme. "Human Evolution: How Random Process Fulfils Divine Purpose." *Perspectives on Science and Christian Faith* 60 (2008) 103–14.
Flanagan, Owen. *Consciousness Reconsidered*. Cambridge: MIT Press, 1992.
Freeman, Thomas B., et al. "Neural Transplantation for the Treatment of Huntington's Disease." *Progress in Brain Research* 127 (2000) 404–11.
Fu, Qiaomei, et al. "An Early Modern Human from Romania with a Recent Neanderthal Ancestor." *Nature* 524 (2015) 216–19.
Fugle, Gary N. *Laying Down Arms to Heal the Creation-Evolution Divide*. Eugene, OR: Wipf and Stock, 2015.

Funk, Cary, and Beka A. Alper. "Religion and Science." Pew Research Center. October 22, 2015. http://www.pewinternet.org/2015/10/22/science-and-religion/.

Funk, Cary, and David Masci. "5 Facts about the Interplay between Religion and Science." Pew Research Center. October 22, 2015. http://www.pewresearch.org/fact-tank/2015/10/22/5-facts-about-the-interplay-between-religion-and-science/.

Funk, Cary, and Lee Rainie. "Americans, Politics and Science Issues." Pew Research Center. July 1, 2015. http://www.pewinternet.org/2015/07/01/americans-politics-and-science-issues/.

Garte, Sy. "Evolution and Imago Dei." *Perspectives on Science and Christian Faith* 65 (2013) 242–44.

Gazzaniga, Michael. "Brain Modularity: Towards a Philosophy of Consciousness." In *Consciousness in Contemporary Science*, edited by A. J. Marcel and E. Besiach, 218–38. Oxford: Oxford University Press, 1988.

George, Andrew. *The Epic of Gilgamesh: The Babylonian Epic Poem and Other Texts in Akkadian and Sumerian*. Translated and with introduction by Andrew George. London: Penguin, 2000.

Goheen, Michael, and Craig Bartholomew. *Living at the Crossroads: An Introduction to Christian Worldview*. Grand Rapids: Baker, 2008.

González, Justo L. *The Story of Christianity*. Vol. 1, *The Early Church to the Dawn of the Reformation*. New York: HarperOne, 2010.

Goodman, Martin. "Paradise, Gardens, and the Afterlife in the First Century CE." In *Paradise in Antiquity: Jewish and Christian Views*, edited by Markus Bockmuehl and Guy G. Stroumsa, 57–63. New York: Cambridge University Press, 2010.

Granqvist, Pehr, et al. "Sensed Presence and Mystical Experiences Are Predicted by Suggestibility, Not by the Application of Transcranial Weak Complex Magnetic Fields." *Neuroscience Letters* 379 (2005) 1–6.

Granqvist, Pehr, and Marcus Larsson. "Contribution of Religiousness in the Prediction and Interpretation of Mystical Experiences in a Sensory Deprivation Context: Activation of Religious Schemas." *Journal of Psychology* 140 (2006) 319–27.

Gray, Harry B., and Jay R. Winkler. "Electron Tunneling through Proteins." *Quarterly Reviews of Biophysics* 36 (2003) 341–72.

Green, Joel B. "'Bodies—That Is, Human Lives': A Re-examination of Human Nature in the Bible." In *Whatever Happened to the Soul?*, edited by Warren S. Brown et al., 149–73. Minneapolis: Fortress, 1998.

———. *Body, Soul, and Human Life: The Nature of Humanity in the Bible*. Grand Rapids: Baker Academic, 2008.

———. "A Response, by Author Joel B. Green, to Scott B. Rae's Review of *Body, Soul, and Human Life: The Nature of Humanity in the Bible*." *Perspectives on Science and Christian Faith* 91 (2009) 194–96.

———. "Restoring the Human Person: New Testament Voices for a Wholistic and Social Anthropology." In *Neuroscience and the Person: Scientific Perspectives on Divine Action*, edited by Robert John Russell et al., 1–22. Vatican City: Vatican Observatory Publications, 1999.

———. "What Does It Mean to Be Human? Another Chapter in the Ongoing Interaction of Science and Scripture." In *From Cells to Souls—and Beyond: Changing Portraits of Human Nature*, edited by Malcolm Jeeves, 179–98. Grand Rapids: Eerdmans, 2004.

Gregory of Nyssa. "On the Soul and the Resurrection." Translated by William Moore and Henry Austin Wilson. In *Nicene and Post-Nicene Fathers*, second series, vol. 5, edited by Philip Schaff and Henry Wace. Buffalo, NY: Christian Literature, 1893. Revised and edited for New Advent by Kevin Knight. http://www.newadvent.org/fathers/2915.htm.

Griffiths, Roland R., et al. "Mystical-Type Experiences Occasioned by Psilocybin Mediate the Attribution of Personal Meaning and Spiritual Significance 14 Months Later." *Journal of Psychopharmacology* 22 (2008) 621–32.

Griffiths, Roland R., et al. "Psilocybin Can Occasion Mystical-Type Experiences Having Substantial and Sustained Personal Meaning and Spiritual Significance." *Psychopharmacology (Berl)* 187 (2006) 268–92.

Griffiths, Roland R., et al. "Psilocybin Occasioned Mystical-Type Experiences: Immediate and Persisting Dose-Related Effects." *Psychopharmacology (Berl)* 218 (2011) 649–65.

Hamer, Dean. *The God Gene: How Faith Is Hardwired into Our Genes*. New York: First Anchor, 2004.

Hameroff, Stuart. "How Quantum Brain Biology Can Rescue Conscious Free Will." *Frontiers in Integrative Neuroscience* 6 (2012) 93.

Hameroff, Stuart, and Roger Penrose. "Consciousness in the Universe: A Review of the 'Orch OR' Theory." *Physics of Life Reviews* 11 (2014) 39–78.

———. "Reply to Criticism of the 'Orch OR Qubit'—'Orchestrated Objective Reduction' Is Scientifically Justified." *Physics of Life Reviews* 11 (2014) 104–12.

Harlow, Daniel C. "After Adam: Reading Genesis in an Age of Evolutionary Science." *Perspectives on Science and Christian Faith* 62 (2010) 179–95.

———. "Creation according to Genesis: Literary Genre, Cultural Context, Theological Truth." *Christian Scholar's Review* 37 (2008) 163–98.

Harris, R Laird. "The Meaning of the Word Sheol as Shown by Parallels in Poetic Texts." *Bulletin of the Evangelical Theological Society* 4 (1961) 129–35.

Hehn, Johannes. "Zum Problem des Geistses im Alten Orient und im Alten Testament." *ZAW* 43 (1925) 216.

Heidel, Alexander. *The Gilgamesh Epic and Old Testament Parallels*. Chicago: University of Chicago, 1946.

Heine, Ronald E. "Origen." In *The Routledge Companion to Early Christian Thought*, edited by D. Jeffrey Bingham, 188–203. London: Routledge, 2010.

Herzog, Hans, et al. "Changed Pattern of Regional Glucose Metabolism during Yoga Meditative Relaxation." *Neuropsychobiology* 23 (1991) 182–7.

Hochschild, Paige E. *Memory in Augustine's Theological Anthropology*. Oxford Early Christian Studies. Oxford: Oxford University Press, 2012.

Hoek, Annewies van den. "Endowed with Reason or Glued to the Senses: Philo's Thoughts on Adam and Eve." In *The Creation of Man and Woman: Interpretations of the Biblical Narratives in Jewish and Christian traditions*, edited by Gerard P. Luttikhuizen, 63–75. Leiden: Brill, 2000.

Irenaeus. *Adversus Haereses*. http://www.newadvent.org/fathers/0103506.htm.

Ishida, Masayoshi. "Rebuttal to Claimed Refutations of Duncan MacDougall's Experiment on Human Weight Change at the Moment of Death." *Journal of Scientific Explanations* 24 (2010) 5–39.

Janssen, Luke J. *Standing on the Shoulders of Giants: Genesis and Human Origins*. Eugene, OR: Wipf & Stock, 2016.

———. "'Fallen' and 'Broken' Reinterpreted in the Light of Evolution Theory." *Perspectives on Science and Christian Faith* 70 (2018) 36–47.
Jeeves, Malcolm. "Brain, Mind, and Behavior." In *Whatever Happened to the Soul?*, edited by Warren S. Brown et al., 73–98. Minneapolis: Fortress, 1998.
———. "Neuroscience, Evolutionary Psychology, and the Image of God." *Perspectives on Science and Christian Faith* 57 (2005) 170–86.
Johnston, Philip S. "Humanity." In *New Dictionary of Biblical Theology*, edited by Desmond Alexander et al., 564–67. Downers Grove: InterVarsity, 2000.
———. *Shades of Sheol: Death and Afterlife in the Old Testament*. Downers Grove: InterVarsity, 2002.
Jones, D. Gareth. "The Emergence of Persons." In *From Cells to Souls—and Beyond: Changing Portraits of Human Nature*, edited by Malcolm Jeeves, 11–33. Grand Rapids: Eerdmans, 2004.
Jonker, Louis C. "Chronicles in an (Un)changing World: The 'Persian Context' in Biblical Studies." *Journal for the Study of the Old Testament* 42 (2018) 267–83.
Josephus. *Antiquities of the Jews*. http://penelope.uchicago.edu/josephus/ant-1.html#EndNote_Ant_1.5a.
Jumper, Chanelle C., and Gregory D. Scholes. "Life—Warm, Wet and Noisy? Comment on 'Consciousness in the Universe: A review of the "Orch OR" theory' by Hameroff and Penrose." *Physics of Life Reviews* 11 (2014) 85–6.
Kim, Seyoon. *Christ and Caesar: The Gospel and the Roman Empire in the Writings of Paul and Luke*. Grand Rapids: Eerdmans, 2008.
Kjaer, T. W., et al. "Increased Dopamine Tone during Meditation-Induced Change of Consciousness." *Brain Research: Cognition and Brain Research* 13 (2002) 255–59.
Korf, Jakob. "Quantum and Multidimensional Explanations in a Neurobiological Context of Mind." *Neuroscientist* 21 (2015) 345–55.
Kosman, Admiel. "Breath, Kiss, and Speech as the Source of the Animation of Life: Ancient Foundations of Rabbinic Homilies on the Giving of the Torah as the Kiss of God." In *Self, Soul and Body in Religious Experience*, edited by A. I. Baumgarten et al., 96–124. Leiden: Brill, 1998.
Kühl, Hjalmar S., et al. "Chimpanzee Accumulative Stone Throwing." *Science Reports* 29 (2016) 22219.
Lamoureux, Denis O. *Evolution: Scripture and Nature Say Yes!* Grand Rapids: Zondervan, 2016.
Landman, Isaac, ed. "Sheol." *Universal Jewish Encyclopedia* 11 (1944) 282–83.
Larson, Jennifer. *Understanding Greek Religion: A Cognitive Approach*. New York: Taylor and Francis, 2016.
Lazar, Sara W., et al. "Functional Brain Mapping of the Relaxation Response and Meditation." *Neuroreport* 11 (2000) 1581–85.
Levenson, Jon D. *Creation and the Persistence of Evil*. Princeton: Princeton University Press, 1988.
Livingstone, David R. *Adam's Ancestors: Race, Religion, and the Politics of Human Origins*. Baltimore: Johns Hopkins University Press, 2008.
Lloyd, Andrew C. *The Anatomy of Neoplatonism*. Oxford: Clarendon, 1990.
Long, Anthony A. *Hellenistic Philosophy: Stoics, Epicureans, Sceptics*. Los Angeles: University of California Press, 1986.
Longman, Tremper, III. *Proverbs*. Grand Rapids: Baker Academic, 2006.

Lou, Hans C., et al. "^{15}O-H2O PET Study of Meditation and the Resting State of Normal Consciousness." *Human Brain Map* 7 (1999) 98–105.

Lynch, Susan V., and Oluf Pedersen. "The Human Intestinal Microbiome in Health and Disease." *New England Journal of Medicine* 375 (2016) 2369–79.

Macaskill, Grant. "Paradise in the New Testament." In *Paradise in Antiquity: Jewish and Christian Views*, edited by Markus Bockmuehl and Guy G. Stroumsa, 64–81. New York: Cambridge University Press, 2010.

MacDonald, Douglas A. "Spirituality: Description, Measurement, and Relation to the Five Factor Model of Personality." *Journal of Personality* 68 (2000) 153–97.

MacDonald, Douglas A., et al. "Spirituality as a Scientific Construct: Testing Its Universality across Cultures and Languages." *Public Library of Science One* 10 (2015) e0117701.

MacDougall, Duncan. "Hypothesis Concerning Soul Substance Together with Experimental Evidence of the Existence of Such Substance." *American Medicine* 2 (1907) 240–43. https://diogenesii.files.wordpress.com/2014/04/duncan_macdougall_1907_-_21pp.pdf.

Madigan, Kevin J., and Jon D. Levenson. *Resurrection: The Power of God for Christians and Jews*. New Haven: Yale University Press, 2008.

Martin-Achard, Robert. *From Death to Life: A Study of the Development of the Doctrine of the Resurrection in the Old Testament*. Edinburgh: Oliver and Boyd, 1960.

Maskie, David. "Scientists and Belief." Pew Research Center. November 5, 2009. http://www.pewforum.org/2009/11/05/scientists-and-belief/.

McDermott, Gerald R. *God's Rivals: Why Has God Allowed Different Religions?* Downers Grove: InterVarsity, 2007.

McFarlane, Graham. Review of *The Human Person in Science and Theology* (London: T. & T. Clark, 2000), by Niels Henrik Gregersen. *Science and Christian Belief* 14 (2002) 94–95.

McGrath, Alister E. *The Science of God: An Introduction to Scientific Theology*. Grand Rapids: Eerdmans, 2004.

McGrath, Gavin Basil. "Soteriology: Adam and the Fall." *Perspectives on Science and Christian Faith* 49 (1997) 252–63.

McGuckin, John Anthony. "Anthropology." In *Westminster Handbook to Patristic Theology*, 13–15. Louisville: Westminster John Knox, 2004.

Middleton, J. Richard. *The Liberating Image: The Imago Dei in Genesis 1*. Grand Rapids: Baker/Brazos, 2005.

Mitchell, David C. "'God Will Redeem My Soul from Sheol': The Psalms of the Sons of Korah." *Journal for the Study of the Old Testament* 30 (2006) 365–84.

Moehlman, Conrad Henry. "The Origin of the Apostles' Creed." *Journal of Religion* 13 (1933) 301–19.

Montgomery, David. *The Rocks Don't Lie: A Geologist Investigates Noah's Flood*. New York: Norton, 2013.

Montgomery, James A. "The Holy City and Gehenna." *Journal of Biblical Literature* 27 (1908) 24–47.

Moreland, J. P., and Scott B. Rae. *Body and Soul: Human Nature & the Crisis in Ethics*. Downers Grove: InterVarsity, 2000.

Murphy, Nancey. "Human Nature: Historical, Scientific, and Religious Issues." In *Whatever Happened to the Soul?*, edited by Warren S. Brown et al., 1–29. Minneapolis: Fortress, 1998.

———. Introduction to *Neuroscience and the Person: Scientific Perspectives on Divine Action*, edited by Robert John Russell et al., i–xxxv. Vatican City: Vatican Observatory Publications, 1999.

———. "Non-reductive Physicalism: Philosophical Challenges." In *Personal Identity in Theological Perspective*, edited by Richard Lints et al., 95–118. Grand Rapids: Eerdman's, 2006.

———. "Non-reductive Physicalism: Philosophical Issues." In *Whatever Happened to the Soul?*, edited by Warren S. Brown et al., 127–48. Minneapolis: Fortress, 1998.

Nagel, Zachary D., and Judith P. Klinman. "Tunneling and Dynamics in Enzymatic Hydride Transfer." *Chemical Reviews* 106 (2006) 3095–118.

Newberg, Andrew B. "The Neuroscientific Study of Spiritual Practices." *Frontiers in Psychology* 18 (2014) 215.

Newberg, Andrew B., et al. "The Measurement of Regional Cerebral Blood Flow during the Complex Cognitive Task of Meditation: A Preliminary SPECT Study." *Psychiatry Research Neuroimaging* 106 (2001) 113–22.

Newberg, Andrew B., and J. Iversen. "The Neural Basis of the Complex Mental Task of Meditation: Neurotransmitter and Neurochemical Considerations." *Medical Hypotheses* 61 (2003) 282–91.

Newport, Frank. "In U.S., 42% Believe Creationist View of Human Origins." Gallup. June 2, 2014. https://news.gallup.com/poll/170822/believe-creationist-view-human-origins.aspx.

New York Times. "Soul Has Weight, Physician Thinks." *New York Times*, March 11, 1907. https://www.nytimes.com/1907/03/11/archives/soul-has-weight-physician-thinks-dr-macdougall-of-haverhill-tells.html.

Nichols, David E. "Psychedelics." *Pharmacological Reviews* 68 (2016) 264–355.

Niehoff, Maren R. "Philo's Scholarly Inquiries into the Story of Paradise." In *Paradise in Antiquity: Jewish and Christian Views*, edited by Markus Bockmuehl and Guy G. Stroumsa, 28–42. New York: Cambridge University Press, 2010.

Noort, Ed. "The Creation of Man and Woman in Biblical and Ancient Near Eastern Traditions." In *The Creation of Man and Woman: Interpretations of the Biblical Narratives in Jewish and Christian traditions*, edited by Gerard P. Luttikhuizen, 1–18. Leiden: Brill, 2000.

———. "Gan-Eden in the Context of the Mythology of the Hebrew Bible." In *Paradise Interpreted: Representations of Biblical Paradise in Judaism and Christianity*, edited by Gerard P. Luttikhuizen, 21–36. Leiden: Brill, 1999.

O'Connell, Robert J. *St. Augustine's Early Theory of Man, A.D. 386–391*. Cambridge: Harvard University Press, 1968.

Oizumi, Masafumi, et al. "From the Phenomenology to the Mechanisms of Consciousness: Integrated Information Theory 3.0." *Public Library of Science Computational Biology* 10 (2014) e1003588.

Pääbo, Svante. *Neanderthal Man: In Search of Lost Genomes*. New York: Basic, 2014.

Pahnke Walter N. *Drugs and Mysticism: An Analysis of the Relationship between Psychedelic Drugs and the Mystical Consciousness*. Boston: Harvard University, 1963.

Panitchayangkoon, G., et al. "Long-Lived Quantum Coherence in Photosynthetic Complexes at Physiological Temperature." *Proceedings of the National Academy of Science* 107 (2010) 12766–70.

Parkes, David. "The Vulnerability of Persons: Religion and Neurology." In *From Cells to Souls—and Beyond: Changing Portraits of Human Nature*, edited by Malcolm Jeeves, 34–57. Grand Rapids: Eerdmans, 2004.

Pasnau, Robert. *Thomas Aquinas on Human Nature*. Cambridge: Cambridge University Press, 2002.

Peacocke, Arthur. "A Christian Materialism?" In *How We Know*, edited by Michael Shafto and Gerald M. Edelman, 146–68. New York: Harper & Row, 1985.

Pedersen, Johannes. *Israel: Its Life and Culture*. London: Oxford University Press. 1946.

Persinger, Michael A. "Religious and Mystical Experiences as Artifacts of Temporal Lobe Function: A General Hypothesis. *Perception and Motor Skills* 57 (1983) 1255–62.

———. "Vectorial Cerebral Hemisphericity as Differential Sources for the Sensed Presence, Mystical Experiences and Religious Conversions." *Perception and Motor Skills* 76 (1993) 915–30.

Persinger, Michael A., et al. "The Electromagnetic Induction of Mystical and Altered States within the Laboratory." *Journal of Consciousness Exploration & Research* 1 (2010) 808–30.

Peschanski, Marc, et al. "Rationale for Intrastriatal Grafting of Striatal Neuroblasts in Patients with Huntington's Disease." *Neuroscience* 68 (1995) 273–85.

Peters, Ted. "Resurrection of the Very Embodied Soul?" In *Neuroscience and the Person: Scientific Perspectives on Divine Action*, edited by Robert John Russell et al., 305–26. Vatican City: Vatican Observatory Publications, 1999.

Pew Research Center. "An Elaboration of AAAS Scientists' Views." July 23, 2015. http://www.pewresearch.org/science/2015/07/23/an-elaboration-of-aaas-scientists-views/.

———. "Evolution and Perceptions of Scientific Consensus." Chapter 4 of *Americans, Politics and Science Issues*. Report. July 1, 2015. https://www.pewresearch.org/science/2015/07/01/chapter-4-evolution-and-perceptions-of-scientific-consensus/.

———. "Perception of Conflict between Science and Religion." October 22, 2015. https://www.pewresearch.org/science/2015/10/22/perception-of-conflict-between-science-and-religion/.

Philo. *Legum Allegoriae*. https://www.loebclassics.com/view/philo_judaeus-allegorical_interpretation_genesis_i_ii/1929/pb_LCL226.139.xml.

———. *Questions and Answers on Genesis*. http://www.earlychristianwritings.com/yonge/book41.html.

Pinker, Aron. "Sheol." *Jewish Bible Quarterly* 23 (1995) 168–79.

Pinkston, William S. *Biology for Christian Schools*. Greensville, SC: Bob Jones University Press, 1991.

Pinnock, Clark H. "Climbing Out of a Swamp: The Evangelical Struggle to Understand the Creation Texts." *Interpretation* 43 (1989) 143–55.

Plato. *Symposium*. http://classics.mit.edu/Plato/symposium.html.

Plotinus. *Enneads*. Translated by A. H. Armstrong. 7 vols. Cambridge: Harvard University Press, 1984.

Preuss, Horst Dietrich. *Old Testament Theology*. 2 vols. Old Testament Library. Louisville: Westminster John Knox, 1992.

Quirke, Stephen. *Exploring Religion in Ancient Egypt*. Chichester, UK: Wiley, 2015.

Reid, Ann, and Shannon Green. "FAQ: Human Microbiome." American Academy of Microbiology. January 2014. http://www.asmscience.org/content/report/faq/faq.3.

Reimers, Jeffrey R., et al. "The Revised Penrose-Hameroff Orchestrated Objective-Reduction Proposal for Human Consciousness Is Not Scientifically Justified." *Physics of Life Reviews* 11 (2014) 101–3.

Rodgers, Christopher T., and P. J. Hore. "Chemical Magnetoreception in Birds: The Radical Pair Mechanism." *Proceedings of the National Academy of Sciences U.S.A.* 106 (2009) 353–60.

Roffman, Itai, et al. "Preparation and Use of Varied Natural Tools for Extractive Foraging by Bonobos (*Pan Paniscus*)." *American Journal of Physical Anthropology* 158 (2015) 78–91.

Routledge, Robin L. "Death and the Afterlife in the Old Testament." *Journal of European Baptist Studies* 9 (2008) 22–39.

Sahu, Satyajit, et al. "Atomic Water Channel Controlling Remarkable Properties of a Single Brain Microtubule: Correlating Single Protein to Its Supramolecular Assembly." *Biosensors and Bioelectronics* 47 (2003) 141–48.

Sahu, Satyajit, et al. "Multi-Level Memory-Switching Properties of a Single Brain Microtubule." *Applied Physics Letters* 102 (2013) 123701.

Sanders, Seth L. *The Invention of Hebrew.* Chicago: University of Illinois Press, 2009.

Scafi, Alessandro. *Maps of Paradise.* University of Chicago Press, 2013.

Schaper, Joahim. "The Messiah in the Garden: John 19.38–41, (Royal) Gardens, and Messianic Concepts." In *Paradise in Antiquity: Jewish and Christian Views*, edited by Markus Bockmuehl and Guy G. Stroumsa, 15–27. New York: Cambridge University Press, 2010.

Scharen, Hans. "Gehenna in the Synoptics." *Bibliotheca Sacra* 155 (1998) 324–37.

Schwartz, Jeffrey M., et al. "Quantum Physics in Neuroscience and Psychology: A Neurophysical Model of Mind-Brain Interaction." *Philosophical Transactions of the Royal Society of London Biological Sciences* 360 (2005) 1309–27.

Scobie, C. H. H. "History of Biblical Theology." In *New Dictionary of Biblical Theology*, edited by T. Desmond Alexander and Brian S. Rosner, 11–20. Downers Grove: InterVarsity, 2000.

Scott, Rodney J., and Raymond E. Phinney Jr. "Relating Body and Soul: Insights from Development and Neurobiology." *Perspectives on Science and Christian Faith* 64 (2012) 90–107.

Scurlock, JoAnne. "Mortal and Immortal Souls, Ghosts and the (Restless) Dead in Ancient Mesopotamia." *Religion Compass* 10 (2016) 77–82.

Shields, Christopher. "Philosophy before Socrates." In *Ancient Philosophy: A Contemporary Introduction*, 1–33. New York: Routledge, 2012.

Shults, F. LeRon. *Reforming Theological Anthropology: After the Philosophical Turn to Relationality.* Grand Rapids: Eerdmans, 2003.

Siemens, David F., Jr. "Neuroscience, Theology, and Unintended Consequences." *Perspectives on Science and Christian Faith* 57 (2005)187–90.

Slifkin, Nathan. *The Challenge of Creation: Judaism's Encounter with Science, Cosmology, and Evolution.* 2nd ed. New York: Zoo Torah / Yashar, 2008.

Stanton, Graham. *The Gospels and Jesus.* 2nd ed. Oxford: Oxford University Press, 2002.

Stearley, Ralph F. "Assessing Evidences for the Evolution of a Human Cognitive Platform for 'Soulish Behaviors.'" *Perspectives on Science and Christian Faith* 61 (2009) 152–74.

Steenberg, M. C. *Of God and Man: Theology as Anthropology from Irenaeus to Athanasius*. New York: T. & T. Clark, 2009.

Stoeger, William R. "The Mind-Brain Problem, the Laws of Nature, and Constitutive Relationships." In *Neuroscience and the Person: Scientific Perspectives on Divine Action*, edited by Robert John Russell et al., 129–46. Vatican City: Vatican Observatory Publications, 1999.

Stone, Lawson G. "Early Israel and its Appearance in Canaan." In *Ancient Israel's History: An Introduction to Issues and Sources*. Grand Rapids: Baker Academic, 2014.

Stroumsa, Guy G. "The Paradise Chronotrope." Introduction to *Paradise in Antiquity: Jewish and Christian Views*, edited by Markus Bockmuehl and Guy G. Stroumsa, 1–14. New York: Cambridge University Press, 2010.

Suriano, Matthew. "Sheol, the Tomb, and the Problem of Postmortem Existence." *Journal of Hebrew Scriptures* 16 (2016) 1–31.

Swinburne, Richard. *The Evolution of the Soul*. Rev. ed. Oxford: Clarendon, 1997.

Tate, W. Randolph. *Biblical Interpretation: An Integrated Approach*. Grand Rapids: Baker, 2008.

Tattersall, Ian. "The Acquisition of Human Uniqueness: How We Got from There to Here, and How We Did It so Fast." In *Human Origins and the Image of God*, edited by Christopher Lilley and Dnaiel J. Pedersen, 25–42. Grand Rapids: Eerdmans, 2017.

Tertulian. *Adversus Marcionem*. http://www.newadvent.org/fathers/0312.htm.

Teugels, L. "The Creation of the Human in Rabbinic Interpretation." In *The Creation of Man and Woman: Interpretations of the Biblical Narratives in Jewish and Christian Traditions*, edited by Gerard P. Luttikhuizen, 107–27. Leiden: Brill, 2000.

Tigchelaar, Eibert J. C. "Eden and Paradise: The Garden Motif in Some Early Jewish Texts (1 Enoch and Other Texts Found at Qumran)." In *Paradise Interpreted: Representations of Biblical Paradise in Judaism and Christianity*, edited by Gerard P. Luttikhuizen, 37–62. Leiden: Brill, 1999.

Toorn, Karel van der. *Family Religion in Babylonia, Syria and Israel: Continuity and Change in the Forms of Religious Life*. Leiden: Brill, 1996.

———. *Scribal Culture and the Making of the Hebrew Bible*. Cambridge: Harvard University Press, 2007.

Torrance, Alan J. "What Is a Person?" In *From Cells to Souls—and Beyond: Changing Portraits of Human Nature*, edited by Malcolm Jeeves, 199–232. Grand Rapids: Eerdmans, 2004.

Truett, K. R., et al. "A Model System for Analysis of Family Resemblance in Extended Kinship of Twins." *Behavior Genetics* 24 (1994) 35–49.

Turin, L. "A Spectroscopic Mechanism for Primary Olfactory Reception." *Chemical Senses* 21 (1996) 773–91.

Turner, Daniel B., et al. "Quantitative Investigations of Quantum Coherence for a Light-Harvesting Protein at Conditions Simulating Photosynthesis." *Physical Chemistry Chemical Physics* 14 (2012) 4857–74.

Turner, Denys. *Thomas Aquinas: A Portrait*. New Haven: Yale University Press, 2013.

Venema, Dennis R. "Genesis and the Genome: Genomics Evidence for Human-Ape Common Ancestry and Ancestral Hominid Population Sizes." *Perspectives on Science and Christian Faith* 62 (2010) 166–78.

Waltke, Bruce K. *The Book of Proverbs: Chapters 1–15*. 2 vols. Grand Rapids: Eerdmans, 2004.

———. *Genesis: A Commentary*. Grand Rapids: Zondervan, 2001.

Walton, John H. *The Lost World of Genesis One*. Downers Grove: InterVarsity, 2009.

———. *The Lost World of Adam and Eve: Genesis 2–3 and the Human Origins Debate*. Downers Grove: IVP Academic, 2015.

Ward, Keith. *The Big Questions in Science and Religion*. West Conshohocken, PA: Templeton Foundation, 2008.

Warren, Matthew. "Mum's a Neanderthal, Dad's a Denisovan: First Discovery of an Ancient-Human Hybrid." *Nature* 560 (2018) 417–18.

Watts, Rikk. "Making Sense of Genesis 1." *Stimulus* 12 (2004) 2–12.

Webb, Taylor W., and Michael S. A. Graziano. "The Attention Schema Theory: A Mechanistic Account of Subjective Awareness." *Frontiers in Psychology* 6 (2015) 500.

Whybray, Roger N. *Proverbs*. The New Century Bible Commentary. Grand Rapids: Eerdmans, 1994.

Wilcox, David L. "A Proposed Model for the Evolutionary Creation of Human Beings: From the Image of God to the Origin of Sin." *Perspectives on Science and Christian Faith* 68 (2016) 2–43.

Wiles, Maurice. "Religious Authority and Divine Action." In *God's Activity in the World: The Contemporary Problem*, edited by Owen C. Thomas, 181–94. Chico, CA: Scholars, 1983.

Williams, Bernard. "The Self and the Future." In *Problems of the Self: Philosophical Papers, 1956–73*. Cambridge: Cambridge University Press, 1973.

Woollett, Katherine, and Eleanor A. Maguire. "Acquiring 'the Knowledge' of London's Layout Drives Structural Brain Changes. *Current Biology* 21 (2011) 2109–14.

Wright, N. T. *The Day the Revolution Began: Reconsidering the Meaning of Jesus's Crucifixion*. New York: HarperCollins, 2016.

———. *Surprised by Hope: Rethinking Heaven, the Resurrection, and the Mission of the Church*. San Francisco: HarperOne, 2008.

———. "The Transforming Reality of the Bodily Resurrection." In *The Meaning of Jesus: Two Visions*, 111–27. New York: HarperCollins, 1999.

Young, Davis A. "The Antiquity and the Unity of the Human Race Revisited." *Christian Scholar's Review* 24 (1995) 380–96.

Young, Frances M., and Andrew Teal. *From Nicea to Chalcedon: A Guide to the Literature and Its Background*. Grand Rapids: Baker Academic, 2010.

Zahnd, Brian. *Sinners in the Hands of a Loving God: The Scandalous Truth of the Very Good News*. New York: Watermark, 2017.

Subject Index

AAAS, 65–68
Aaron, 78, 141, 142
Abel, 18–19
Abimelech, 5–6
Abraham, Abram, *xii*, 3–6, 10, 14, 17, 38, 42, 133, 142–44, 163, 164, 174, 184–85
abstract thinking, see cognitive function
Abusch, Tzvi, 39
abyss, 137, 138; also see Sheol
Accuser, see Satan
Adam, 161
    Adam and Eve, 17, 39, 52, 53, 55, 68, 77, 121, 125, 157, 158, 165–67, 186, 187, 194
    historicity, 68, 69
    Second Adam and Eve, 53
    also see human origins—from clay
afterlife, 129–83
    ancient Hebrew view, 135–44
    ancient Mesopotamian view, 131–32, 191–92
    Elysium, Elysian Fields, 38, 193
    Egyptian Field of Reeds, 36, 132
    Hades, 38, 144, 155, 157, 183, 186, 193
    Heaven, 31, 44, 52, 62, 79, 88, 127, 130, 139, 145, 150, 157–58, 163–68, 172–77, 180–81, 186–89, 191
    Hell, 52, 77, 82, 127, 155–57, 182, 186; also see Gehenna
    Intermediate State, 56, 173–74, 183, 185
    no afterlife at all, 146
    Paradeisos, 29, 163, 183
    Paradise, 55, 133, 139, 144, 146, 147, 150, 157–68, 174–75, 182, 186, 193
    prehistoric view, 130–31; also see ritual burial
    Purgatory, 165, 173–75, 183, 185
    Sheol, see Sheol
    Tartaros, 38, 157
    underworld, underground, 81, 131, 137, 193
    Zoroastrian view, 133, 146, 182, 191
aging, see brain dysfunction
Ahaz, 155
Akh, 132
Akkad, 4
    –ian Empire, 4, 7, 16, 35
    –ian culture, 4, 7
    –ian mythology, see mythology—Akkadian
    –ian texts, 7, 131, 137
Alexa, 92, 109, 113
Alexander the Great, 12, 163
allegory, see writing styles
Alzheimer's, see brain dysfunction
American Association for the Advancement of Science, 65-68
Ammut, 36
analogy, see writing styles
ancestor worship, 38, 79, 133–34

SUBJECT INDEX

Ancient Near East(ern), *xii*, *xv*, 3, 4, 7, 11, 16, 19, 21–25, 32, 34, 38, 41, 78, 126, 133, 147–48, 156, 185
Anderson, Ray, 61
angel(s), 5, 12, 44, 50, 55, 80, 86, 106, 126, 147, 151, 158, 163, 173, 176, 186
anthropogony, see human origins
Apostles' Creed, 155, 177
Appolinaris of Laodicea, 55
Aquinas, Thomas, 58, 59, 62, 123
Aramaic, 7, 27, 137, 169, 170
Archangel, Michael, *iv*, 1
Aristotle, Aristotelianism, 12, 13, 37, 46–49, 56–59, 62, 88
Artaxerxes, 12
artificial intelligence, 92, 102, 108, 109, 113, 199; also see: Alexa; Cortana; Google; Siri
Asherah, 8
Assyrian empire, 3, 161–62
Athanasius of Alexandria, 55, 85
Athena, 16
Athenagoras, 155, 179
atom, 47, 179, 180
Attention Schema Theory, see consciousness
Auffarth, Christoph, 165
Augustine, 1, 13, 24, 53, 56–58, 85; 111, 123, 124, 179–80
authority, see Scripture
axon, see cognitive function—synapse

Babylonian writing, 27
Baal, 8
bacteria, 115, 119, 167, 178–80
baptism, 30, 191
Bar, Shaul, 137, 140
Barna Group, 67, 70, 195, 196
Barnabas, 192
Barth, Karl, 62
Bartholomew, Craig, 69, 113, 197
basar, 42, 79, 80
Bauckham, Richard, 160
bench-tombs, 17, 143
Benjamin, 142
Benjamins, Hendrik S., 165
Bernard of Clairvaux, 165–66

big bang, *xiv*, 66, 68, 114, 194
Bjork, Russell C., 123–24
Bloch-Smith, Elizabeth, 134
blood
  of deity, 15, 16, 35, 39, 126
  human, 34, 41, 59, 115, 116, 180, 199
  equals life, 38, 39
  sacrificial, 38, 39, 133
body-soul dualism, see Substance Dualism
Boers, Hendrikus, 149
Boethius, 58
bones, 17, 131, 134, 137, 143, 144, 146; also see Valley of Dry Bones
brain
  brain regions, 63, 65, 91, 103, 104, 106, 122
  role in cognitive function, *vii*, 48, 59, 65, 72–74, 93–96, 100–104, 109, 113, 118, 198
  cools the blood, 59
  defines self, 33, 48, 87–91, 93, 96, 106, 108, 110, 113, 115, 121, 123, 126, 173, 198
  exerts hydraulic pressure, 61, 85
  motor control, 61, 85
  not residence of soul/consciousness, 33, 34, 36, 40, 41
brain dysfunction
  aging, 95, 112
  Alzheimer's, 94, 112, 180, 198
  beneficial effect of exercise/use, 95
  dementia, 198
  epilepsy, 106
  Huntington's, 87, 198
  injury, 94; also see Gage, Phineas
  locked-in syndrome, 198
  neurodegenerative, 94–95
  Parkinson's, 198
  pavor nocturnis, 199
  Schizophrenia, 94, 180
  split-brain, 94
  tumor, 94
Branson, R. D., 40
breath
  divine B—, 20, 36–37, 40, 53

metaphor for the divine, 14, 16, 37, 39, 40, 53, 125, 126
metaphor for life or spirit, 14–16, 20, 32–33, 38, 39, 53, 145
B— of life, 15, 16, 20, 37, 39, 77, 78, 81, 113, 124
breathed his last, 39, 142
Bremmer, Jan N., 162–63
Brown, Warren S., 94, 110–13, 120–21
Bultmann, Rudolf, 62
Buridan's ass, 119

Caesar Augustus, 170-172
Cain, 18–19
Calvin, John, 197
Canaan, C—ites, 4, 5, 7, 8, 13, 42, 133, 155, 157, 158
cannibal problem, 179, 188
canonization, see writing of the Bible
captivity, see Israel—captivity
Cartesian theater, see consciousness
Catholic, Roman, 28, 66—68, 175, 197
chain consumption, see cannibal problem
chaos, chaotic forces, 14, 17, 21, 119, 143, 194
Charlesworth, James H., 145, 146, 181
chimpanzees, 92, 93, 108, 114, 194; also see: Kanzi; Koko
Chiron, 18
Christian doctrine, Christian tradition
    age of accountability, 191
    ascetism; body is evil, 51, 55, 84, 152, 186; also see Greek philosophy—matter is evil
    atonement, 122, 128, 165
    divinization of Jesus, 52, 125
    Eschaton, 89, 146–47, 155–57
    Fall (in the Garden), 26, 31, 43, 55–57, 77, 128, 161, 166, 187
    immaculate conception, 44
    inerrancy, infallibility; see Scripture—inerrancy, infallibility
    influenced by education, 66–67
    influenced by Greek philosophy, 13, 23, 32, 53, 75, 81, 84, 126–27, 152, 155, 179, 184, 186–87, 192, 197
    influenced by modern science, 61, 62, 66
    influenced by modern philosophy, 88
    influenced by mythology, 126, 156
    influenced by politics, age, ethnicity, gender, 66
    influenced by science, 58, 61–62, 66–70, 85–87, 90, 114–15, 120, 127, 193–94
    introduction of mortality, 68, 77, 124, 135, 166
    Moral Influence Theory of atonement, 122
    original sin, 18, 43, 53, 54, 56, 77, 135, 166, 187
    penal substitution, 175, 187
    resurrection, 11, 25, 44, 53, 56, 79, 81, 84, 117, 129, 133, 136, 145–53, 156, 173, 176–82, 190; also see disembodied existence—re-embodied existence; afterlife
    theosis, 53
    three-tiered cosmology, 68, 103, 158, 164
    Trinity, 49, 52, 57, 186
    Virgin birth, 29, 44
    also see: Apostle's Creed; Catholic; Nicene Creed; Protestant
Chrysostom, John, 55
church fathers, see Fathers of the church
circumcision (circumcised; uncircumcised), 5, 139, 144, 191, 192
clay, see human origins —from clay
Clayton, Philip, 90
Clement of Alexandria, 13, 23, 51, 190, 197
cloning, 87, 198, 200
cognitive function, 40, 41, 52, 61, 63, 75, 91, 92, 94, 96, 111, 118, 123
    abstract thinking, 91, 93, 94, 197
    creativity, 52, 75, 92, 93, 111, 118–19
    dreaming, 38, 43, 79, 91, 95, 105, 199

cognitive function (*continued*)
    emotion, 41, 55, 56, 65, 73, 75, 85, 92, 93, 104, 109, 110, 111, 114, 120
    facial recognition, 91, 109
    free will, see volition
    language, 75, 91, 93, 111, 120
    logic, 110
    meditation, 91, 103, 165
    memory, 42, 57, 65, 85, 93, 94, 100, 110, 111, 116–17, 120
    motor control, motor cortex, 48, 61–63, 85, 91, 94, 108, 110
    neuronal firing, 65, 72–73, 85–86, 91, 93, 97–100, 109, 118, 199
    reason, reasoning, 52, 53, 56, 57, 75, 85, 110, 111
    self-awareness, self-identity, 65, 73–74, 92, 111
    signal processing, 94–95, 108–110
    synapse, synaptic, 100, 103
    tool-making, 93
    vision, visual cortex, 60, 71–73, 91, 109
    volition, 42, 52–54, 56, 65, 75, 85, 89, 91–93, 110, 111, 114, 118–20, 128
Collins, John J., 146, 149
Columbus, Christopher, 160
consciousness, *vii*, 31, 58, 60, 61, 65, 66, 83, 89, 90, 95–102, 105, 110, 119, 124, 127–28
    Attention Schema Theory, 96
    Cartesian theater, 60–61, 96
    emergent property, 111, 115, 119
    Global Workspace Theory, 96
    Integrated Information Theory, 96
    multiple Drafts Model, 96
    Orch-OR model, 101–3, 119
    physicalist explanation, 95–122
    universal consciousness, *viii*
Constantine, 56
Copernicus, Copernican 68, 69, 114
Cortana, 92
Cortez, Marc, 88–89
Council of Jerusalem, 192
creativity, see cognitive function—creativity

Crenshaw, James L., 138
Crick, Francis, 89, 90
cyberspace, 71–73, 90
Cyril of Alexandria, 55
Cyril of Jerusalem, 165
Cyrus the Great, 9, 11, 24

Daniel (the person), 11–12, 24, 81, 84, 133, 145–50, 153, 165
Darwin, Charles, 194
David (King), 10, 40, 42–43, 80, 134–36, 142
death, die(died), 43, 63, 121, 131, 144; also see mortal(—ity)
    "breathed his last," 39, 142
    "gathered to his people," "slept with his fathers," 78, 84, 142, 143
    "gave up his spirit," 82
    introduction of death, see Christian doctrine—introduction of mortality
death cult, 38, 78, 79, 84, 133–35, 143
Deborah, 8
dementia, see brain dysfunction
demons, 186, 199
Denisovans, 114, 124-125
Dennett, Daniel, 60, 96, 101
    Multiple Drafts Model, see consciousness
Descartes, René, 55, 59, 60, 61, 85–86; 111, 124, 126, 201; also see *res cogitans, res extensa*
determinism, 118–19
Deucalion, 16
disembodied existence, 43–44, 47, 78, 79, 80, 84, 89, 117, 125–26, 132, 135, 174–74
    re-embodied existence, 53, 84, 89, 115–17, 148, 152–55, 173, 175–82, 191; also see Christian doctrine—resurrection
divinization of Jesus, see Christian doctrine
divine inspiration, 13, 20, 22, 23, 26, 27, 29, 30, 191–93, 197; see Revelation
DMT, 103
dopamine receptor, 107

SUBJECT INDEX 223

Dorcas, 146
dreaming, see cognitive function
dualism, see substance dualism;
    property dualism; monism
    variety of forms, 88, 128
dust, 116
    used in creating humans, 16, 49, 77, 135
    associated with death, Sheol, 16, 17, 20, 44, 77, 80, 83, 131, 132, 135, 136, 138, 145, 147, 161, 182, 185

Eden, see Garden—of Eden
ego, 61
Egyptian empire, 4, 6, 16, 131
Egyptian mythology, see mythology, Egyptian
Einstein, Albert, 71, 181
Ekur, 156-57
Eli, 8
Elijah, 8, 44, 79, 139, 151, 158, 165
Elisha, 141, 146
Elledge, C. D., 148, 153
Elohim, 5, 44, 49, 161, 162, 186
Elysium, Elysian Fields, 38, 193
embalming, 132
emergence, xii, 74, 88, 90, 91, 96, 101, 111, 115, 119-25, 128, 130, 173
    emergent property, see soul—nature of
emotion(s), see cognitive function
Endsjø, Dag Øistein, 152-53
Enki, 15
Enkidu, 17, 26, 36, 131-32
Enlil, 14, 19, 156, 157
Enoch, 78, 139, 158, 163-65
entanglement, 65
Epicurus, 47
epilepsy, see brain dysfunction
epithymetikon, 47, 57; also see soul—tripartite nature
Ereshkigal, 132, 137
eretz, 137
Eschaton, 89
Essenes, 150
euangelion
    of Rome, 170-73
    of Jesus, 151, 168-73, 192

Euphrates River, 21, 159, 160
Eutyches, 58
Eutychus, 146
Evangelical(ism), 62, 67, 68
Eve, 26, 39, 49, 166; see also Adam—Adam and Eve; human—creation of woman
Exile (Neo-Babylonian), 7, 8, 79, 81, 148
ex nihilo, 44
Ezekiel's Vision of Valley of Dry Bones, see Valley of Dry Bones

facial recognition, see cognitive function
fall (in the Garden), see Christian doctrine
Fathers of the church, 13, 24, 49, 51, 52, 58
feather of Ma'at, 1
final judgement, 1, 77, 129, 132, 139-41, 144, 145, 147, 154-57, 165, 174, 182-83, 185; also see Jesus—*Parousia*; *Eschaton*
fire, flames, 5, 16, 141, 147, 155, 156, 161, 168, 174, 182 185; also see Gehenna
flood (global F—, or Noah's F—), 19, 20, 21, 45, 193, 194
Forms and Ideas, see Platonism
free will, see cognitive function—volition
fungi, 179, 180

Gage, Phineas, 34, 93
Galen, 48
Galileo, 68-70, 113, 114, 197
Gallup Inc., 67
Ganges River, 160
garden, 17, 21, 29, 49, 55, 167
    Gan-Eden, 158, 160, 161
    of Eden, 50, 139, 144, 147, 157-59, 161-64, 166-67, 182, 186, 187
Gehenna, Gehinnom, 79, 155-57, 186
genetics of spirituality, 106-107
giants, see Nephilim
Gideon, 8, 142
Gihon River, 159, 160
Gilgamesh, 1, 36, 131, 194

Global Workspace Theory, see consciousness
Gnosticism, 84, 186
Goheen, Michael, 69, 113, 197
Good Friday Experiment, 104–5
Google, 92, 113
Gospel, see *euangelion*—of Jesus
grave, 137–39, 143; also see Sheol
Greek Empire, 3, 12, 16, 163; also see Hellenism
Greek philosophy
    creative force, the One, the All, the Good, 47–48, 50
    matter is evil, 51, 55, 152, 186
    the *nous*, 46–48, 50, 53, 55, 81, 95
    Logos, 12, 47, 53, 55
    reversion, 48, 53
    soul is part of the universal mind, 56
    soul trapped in body, 46, 52, 54–56, 62, 81, 84
    also see: Aristotelianism; Platonism; Substance dualism
Green, Joel B., 61, 201
Gregory of Nyssa, 54, 179
gut, see soul—location

$5HT_{2A}$ receptors, see serotonin receptors
Hades, see afterlife
Hephaestus, 18
hair, hair length, 13, 23, 30, 40, 115, 134, 192
hallucinogen, 65, 103, 104, 107, 122, 198
Hamer, Dean, 107
Hameroff, Stuart, 96, 102, 119
    Orch-OR model, see consciousness
Hannah, 141
Havilah, 159, 160
heart, see soul—location
heaven, see afterlife
Hebrew script, 7, 28
Heisenberg uncertainty principle, 119
heliocentrism, 34, 68–69; 114
hell, see Afterlife
Hellenism, 12, 13, 23, 31, 43, 47, 54, 58, 81, 145, 148–50, 152, 163–64, 183, 186, 187, 193
helmet (God H—), 106
Heracles, 18

heresy, 51, 84
hermeneutics, see Scripture
Hesiod, 15
Hezekiah, King, 133, 135
Hilkiah, 10
Hippocrates, 48
holism, 42, 56, 59, 76, 77, 126
Holy Spirit, 13, 23, 29, 38, 40, 53, 54, 125, 197
hominid evolution, see human origins—biological evolution
human origins, 1, 4–6
    Ancient Hebrew understanding, 38–44
    Ancient Near Eastern mythology, 15–18, 35–37
    biological evolution, *xi*, 3, 65–67, 68, 124, 127, 188, 194–95
    breath of life, see breath of life
    creation of woman, 39, 49, 77, 83
    created male and female, 49, 50, 55, 111, 177, 178
    created androgynous, non-sexual, immortal, 50, 52, 53, 54, 55
    from clay, dirt, dust, mud, 4, 14–17, 21, 35, 39, 40, 49, 52, 65, 77, 80, 81, 83, 126–27, 135, 184–85
    Greek mythology, 37–38, 45–46
    knit in mother's womb, *xiv*, 16, 39, 81, 118
    migration out of Africa, *viii*, *xi*, *xii*, 3–4, 184
    974 generations existed before Adam, 51
human sacrifice, 8, 134, 155

Ignatius, 13
image of God (*imago Dei*), *viii*, *xi*, 17, 21, 50, 5–55, 69, 75, 111, 113, 124, 167, 185, 191
    Hebrew versus Greek views, 52, 75, 111, 185
immaculate conception, see Christian doctrine
immaterial, 35, 37, 42, 44–48, 51, 52, 54, 55, 58, 60, 71–74, 78, 81, 84, 90, 91, 111, 120, 125

SUBJECT INDEX

influencing the material, see Substance Dualism—mind-body problem
immortal (—ity), 1, 19, 35, 36, 43, 49, 50, 52–54, 61, 77, 115, 117, 129, 146, 147, 149, 152, 153, 164, 200; also see mortal, mortality
incarnation, 125
indulgences, 175
inerrancy, infallibility, see Scripture
Intermediate State, see Afterlife
Integrated Information Theory, see consciousness
intestines, see soul—location
Irenaeus, 13, 23, 53, 54, 165
Isaac, 5, 142, 143
Ishmael, 5, 78, 142
Israel
    captivity, Assyrian, 3
    captivity in Egypt, 3, 5–7, 13, 79
    captivity, neo-Babylonian, 3, 7–9, 12, 13, 147, 149, 155
    editing of sacred texts, 9–11
    evolution of nation, politics, religion, 8, 9, 13, 38
    exile, 7, 8, 79, 81, 148, 149, 156
    exile, Persian, 3, 145–46, 176
    exodus from Egypt, 7, 13, 141, 143
    idol worship, 8, 160, 192
    journey through desert, 5, 13, 133
    monarchy, 8
    post-exile, 11, 25, 81, 148
    receiving the law, 156, 189
    subjugation, under Greece, 3, 163
    subjugation, under Rome, 3, 12, 163, 168, 170

Jacob, 5, 42, 78, 142, 143
James (brother of Jesus), 82, 153, 156
Jehoiakim, 142
Jehoram, 142
Jepthah, 8
Jerome, 56
Jesus, 53, 150–53, 158, 166, 167
    Ascension, 150
    crucifixion, 154
    empty tomb, 150
    firstborn, first-fruit, 153

resurrection, 150, 154
resurrection body not physical, 151, 176
resurrection body physical, 150, 153, 176
Parousia, 54, 55, 172–74
Joab, 136
Job, 135, 136, 140, 144–45
John, the disciple, 125, 151, 169, 176
John, the Revelator, 27, 157, 163–64, 172, 174, 181
John the Baptist, 168
Johnston, Philip, 129, 136, 137, 140, 146–48
Jones, D. Gareth, 197
Joseph, 142, 143
Josephus, 50, 160, 164
Josiah, 11, 133, 142
Joshua, 8
Jotham, 142
Judaism
    evolution of, 77–81
    rabbinical, 13, 50, 157
    Second Temple J—, 9, 12, 13, 51, 144–55, 164
    Hellenistic, 48–51, 53, 54, 81, 149–58
Justin Martyr, 23, 25, 190

Kanzi, the chimpanzee, 113
Khnum, 15
kidneys, see soul—location
Kingdom of Heaven, 167–68, 176, 189
Koko, the gorilla, 93, 113
Korah, sons of Korah, 139–41
Koren, Stanley, 106
kubu, 16, 39, 81

Lazarus, 146, 174
Leah, 143
lebh, 42, 43, 80
lens, *xi*, 13, 22, 57, 190, 193
Leontius of Byzantium, 58
Levantine corridor, *viii*, 3–4
lichens, 74
logistikon, 47, 57; also see soul—tripartite nature
Logos, see Greek philosophy

logos spermatikos, see Revelation
Lot, 5, 162
Lucifer, 161, 187
Luke, the Biblical author 82, 144, 151, 172, 174
Luther, Martin, 30
LSD, 103

Ma'at, 63, 132; also see feather of Ma'at
MacDougall, Duncan, 1, 63–64
maggots, 131, 138, 182
Magi, 24
male, see human origins—created male
Manasseh, 155
Manicheanism, 56, 186
Marcion, 51, 52
Martin-Achard, Robert, 129
martyrdom, 25, 82, 84, 145, 146, 149, 153, 154, 165, 190
Mary, mother of Jesus, 48, 53, 181
Mary Magdalene, 150, 151, 181
Masoretic text, 10, 160
materialism, see physicalism
Medium of Endor, 5, 44, 84, 126, 134
Melchizedek, 24
meditation, see cognitive function
memory, see cognitive function
mescaline, 103
metaphor, see writing styles
microtubules, 97–102
mind-body problem, see Substance Dualism
Mitchell, David C, 141
Molech, see human sacrifice
monism, 42, 62, 74, 76, 77, 83, 89, 118, 128, 185, 200
    variety of forms, 128
monotheism, 12, 52, 133
morality, 75, 93, 200
Moreland, J. P., 173
mortal, mortality, 16, 36, 43, 45, 50, 52, 54, 81, 117, 127, 132, 174; also see immortal(—ity)
Moses, 7, 10, 51, 78, 133, 141, 142, 151, 156, 160
    author of Pentateuch, Torah, 7, 10, 133
    author of *Urdeuteronomium*, 10

Egyptian education, 7, 133
leading Israel, see: Israel—journey; Israel—exodus
motor control, motor cortex, see cognitive function
mountain
    Babylonian holy mountain, 156–57
    Mount Horeb, 156
    Mountain of God, 158, 161, 167
    Mount Nebo, 151
    Mount Sinai, 156, 189
    Mount Zion, 156, 162
multiverse theory, 89
murmuration, 90, 91, 93, 119
Murphy, Nancey, 61, 87, 120, 179, 185
mythology
    Babylonian, 14–20, 26, 35, 126, 137, 159, 184, 190
    Egyptian, 14–16, 36, 37, 63, 184
    Greek, 15–16, 18–20, 26, 37, 152

Naaman, 24
Nathan, 8
naturalism, see physicalism
Neanderthals, 107, 114, 124–25, 194
Nebuchadnezzar, 9
Nehemiah, 12
Nemesius of Emesia, 56
Neolithic, 138
Neoplatonism, 13, 46, 47, 51, 53, 56, 58, 62, 92; also see Platonism
nephesh, 40, 42–44, 50, 54, 77, 79, 80, 83, 113, 185
Nephilim, 20, 44, 45
Nergal, 132
Nestorius, 58
neurotransmitters, 65, 91, 92, 103, 107
neuronal firing, see cognitive function
New Jerusalem, 158, 163, 164, 165, 174, 181
Newton, Isaac, 86
Nicene Creed, 155
Nicodemus, 121
Nile River, 21, 160
Nintu, 15
Noah, 19, 45; also see flood
non-canonical books, 29–30

SUBJECT INDEX

non-reductive physicalism, see physicalism
Nous, see Greek philosophy

old earth, 70
ontology, 1, 4, 7, 31, 38–64, 79, 81, 184, 189
oral tradition, 6, 11, 15, 26, 27, 30
Optatus, 123
Orch-OR, see consciousness
Origen, 23, 53, 155
original sin, see Christian doctrine
Orthodox church, 28, 42, 175
orthodoxy, 51
oxytocin receptor, 107

Paleolithic, 138
Pandora, 16, 18, 19, 26
Paradise, see afterlife
parallelism, see writing styles
Parousia, see Jesus
Pasnau, Robert, 58
Patristic, 22, 31–32, 43, 46, 49, 51, 55, 58, 75, 127; also see fathers of the church
Paul, Apostle, 23, 24, 30, 31, 40, 48, 51, 54, 75, 82, 84, 103, 122, 123, 149, 150, 152, 153, 155, 156, 163, 164, 167, 172, 174, 177, 178, 181, 187, 190, 192
Paullus Fabius Maximus, Proconsul, 170–71
pax Romana, 171
Penrose, Sir Roger, 96, 102, 119
   Orch-OR model, see consciousness
Pentateuch, see Moses, author
persecution, 25, 146–49, 153, 154, 165, 171, 190; also see martyrdom
Persian Empire, 11–13, 133, 146, 162–63, 182, 184
Persinger, Michael, 106
personality, xii, 33, 34, 36, 40, 73, 74, 92–95, 103, 106, 117, 118, 120–22, 125, 126, 185, 199, 200
Peter, the disciple, 174, 177, 192
Peters, Ted, 179
Pew Research, 65–68

Pharisees, 150, 177
Philo, 48, 50, 52, 81, 159
Phinney, Raymond E., 87, 89, 123–24, 179
phrenology, 63, 195
physicalism, 58, 85, 89, 90, 95, 103
   non-reductionist P—, 90, 108, 110, 113, 120, 127
   challenges raised against, 113–19
   P— explanation of consciousness, see consciousness
   P— explanation of human origins, see human origins—biological evolution
   P— explanation of soul, 108, 110–13
   P— explanation of spirituality, 103, 106–7, 120–22
pineal gland, 61, 85
Pishon River, 159, 160
pit, 38, 137, 138, 141; also see Sheol
placebo, placebo effect, 104; 111–12
plagiarism, see writing styles
Plato, 12, 37, 44, 46, 47, 48, 49, 51, 57, 62, 123, 187
Platonism, 12, 13, 46, 47, 49, 50, 51, 56, 62, 187; also see Neoplatonism
   Forms, Ideas, 46, 47, 50, 56, 57, 58, 62, 88, 123, 180
Plotinus, 47
Plutarch, 187
pneuma, 31, 38, 40, 117
polemics, see writing styles
Polycarp, 13
progressive revelation, see Revelation
Prometheus, 16, 18, 19
property dualism, 58, 89
Protestant, 28, 30, 66–68, 175
psilocybin, 103–5
psyche, 31, 37, 38, 117
psychedelic drugs, see hallucinogen
psychosomatic unity (or whole), 56, 58, 74, 76, 84, 88, 89
Ptah, 15
Purgatory, see Afterlife
Pyrrah, 16
Pythagoras, 12

quantum mechanics, 65, 89, 97, 98, 101–3, 118–20, 127, 151, 194; also see: entanglement; Heisenberg; Schrödingers; string theory; superposition
Qoheleth, 138

Ra-Atum, 15
Rae, Scott B., 173
reason, see cognitive function
Rebekkah, 143
receptor (neural), 91, 103, 105; also see: dopamine; oxytocin; serotonin
reductionism, 90; also see physicalism—non-reductive
Reformation, 61, 175; also see Luther
Renaissance, 48, 166, 193
rephaim, 81, 84, 136, 138, 148, 173
res cogitans, 59–60
res extensa, 59–60
resurrection, see Christian doctrine
retina, 60, 72, 73, 109; also see cognitive function—vision
revelation
    General R—, 190, 191
    Progressive R—, 23, 25, 154, 189–91, 196–97
    logos spermatikos, 23, 25, 190, 191
    to pagans, 51
    also see: Divine inspiration; theophany; writing of the Bible
ritual burial
    prehistoric, 107, 130
    Egyptian, 36, 131
    Hebrew, 142–43
    Tibetan, 131
Robinson, John A. T., 62
Roman Catholic, see Catholic
Roman Empire, 3, 12, 56, 163, 172
Routledge, Robin L., 154
ruach, 40, 42–44, 80

Sadducees, 150, 177
Sagan, Carl, 65
Samson, 8, 40
Samuel, 8, 78, 141
    spirit raised by medium of Endor, 5, 44, 84, 126, 134
Sarah, 143

sarx, 31, 53, 54
Satan, 12, 80, 82, 126, 186
Saul, King, 5, 9, 44, 84, 126, 134
Scafi, Allesandro, 162
Scientific Revolution, 166
schizophrenia, see brain dysfunction
Schrödingers cat, 101
Scripture
    authority, 68, 115
    authorship, see: Divine inspiration; Moses; writing of the Bible
    reflects only the science of their day, 22, 33–34, 41, 56, 103, 114, 197
    hermeneutics, 29, 154, 193
    inerrancy, infallibility, 28, 195
    interpretation influenced by science, 70, 166, 193–94
    also see Word of God
Scott, Rodney J, 87, 89, 123–24, 179
Second Temple Judaism, see Judaism, Second Temple
Second Coming of Christ, see Jesus—*Parousia*
sensory input, 61, 62, 96, 103, 105
serotonin, 65, 103
serotonin receptors ($5HT_{2A}$-receptors), 103—105, 107, 198
Shema, 186
Sheol, 44, 78–81, 80, 83, 84, 129, 135–48, 154–57, 164, 167, 173, 175, 182, 190
    abandoned, cut off, forgotten, 135, 137, 140–42, 147, 182
    etymology, 136–37
    place of punishment, 141, 157, 175
    place of rest, not of punishment, 136, 148, 155, 167182
    return from Sheol impossible, 38, 44, 84, 129, 137, 182
    return from Sheol successful, 80, 137, 141
    like a prison, 83, 132, 138
    gloomy, non-existence 38, 52, 84, 129, 146, 182
    universal destiny, 138–39, 144, 164, 190
    also see: abyss; afterlife; dust; maggots; pit; *rephaim*
Shibboleth, 28, 195

SUBJECT INDEX

Shunnamite woman, 146
Siemens, David F., 125
signal processing, see cognitive function
Siri, 92, 109, 113
Sisyphus, 86
Socrates, Socratism, 12, 37, 46
Solomon, 136, 142
sons of God, of Shem, of Ham, of Cain, see *Nephilim*
soul,
    disembodied, see disembodied
    location in body:
        chest, not head, 33, 34, 36, 40
        gut, intestines, 34, 40, 41, 114
        head, brain, 33, 48, 59–63, 87–88, 91–103
        heart, 34, 41, 42, 48, 59, 114
        kidney, 41, 114
        liver, 40
        whole body, 46, 59, 123–24
    nature of:
        distinct from spirit, 48–49, 82, 121
        emergent property, 74, 88, 90, 96, 108, 111–13, 121, 124, 127
        trapped in body, see Greek philosophy
        tripartite, 35, 46–47, 49, 53, 57, 59, 62, 185
        weight, 1, 63–64, 132
    origin of:
        assigned to a given body, 39, 54, 59, 123
        pre-exists conception, 39, 53, 54, 57, 87, 123–24
        created at conception, 56, 57, 87, 89, 110–13, 117, 120–21, 123–24
        Traducianism, 57, 123
        develops/matures during life, 54, 89
        ceases to exist at death, 47, 89
    understanding of:
        ancient Hebrew view, 39–44
        Babylonian view, 35–36
        Egyptian view, 36–37
        Mesopotamian view, 35–36
        Greek view, 46–48
        modern non-reductive physicalist view, 108–113

    also see: *pneuma*; *psyche*; *ruach*; *nephesh*; soulishness; spirit
soulishness, 110, 113, 120
spirit (human), 42, 43; also see Holy Spirit
    different from soul, 121
    disembodied, see disembodied existence
    transferred to another, 79
    travelling outside body, 79, 80
spirituality, physicalist explanations, 106–7, 120–22
split-brain, see brain dysfunction
stars, shining like S—, 146, 147, 151, 153
Stephen, 144
Stoicism, 24, 47, 48, 51, 53, 55, 56
string theory, 89
substance dualism, 35, 42, 50, 55, 58, 60, 62, 74, 76, 84, 89
    arguments against, 85–87, 200
    biblical or not, 76–85, 114
    division between laypeople and scholars, 62, 84–85, 88, 127, 200–201
    mind-body problem, 55–56, 60–61, 85–87, 111, 113, 126, 128, 188, 201
Sumer, 4, 159
    –ian empire, 4, 6, 16, 21, 35
    –ian culture, 4, 6, 135
superposition, 65, 101, 119
Suriano, Matthew, 142

Tabitha, 146
Talmud, 51
Tartaros, see afterlife
Tertullian, 54, 89, 123, 155, 165
theophany, 5, 120
theosis, see Christian doctrine
Thomas (the disciple of Jesus), 150
Thomism, see Aquinas
thymikon, 47, 57; also see soul, tripartite nature
Tigris River, 159, 160
Torah, 4, 7, 10–11, 40, 77–79; also see: Moses; Pentateuch
Torah, see Moses, author
Traducianism, see soul—Traducianism

SUBJECT INDEX

transhumanism, 117, 199
Tree of Knowledge of Good and Evil, 16, 26, 77, 121, 159
Tree of Life, 16, 43, 159
trees of Paradise, 159, 161–62, 167
Trinity, see Christian doctrine
tubulin, 65, 96–102, 106
twins, 87, 115, 124
Tyre, king of, 160–62, 187

Ur of the Chaldeans, 4
Ut-napishtim, 19, 35, 36

Valley of Dry Bones (Ezekiel's vision), 39–40, 44, 79, 147, 148
Valley of Hinnom, see Gehenna
van der Toorn, Karl, 8–11
Vespucci, Amerigo, 160
virgin birth, see Christian doctrine
virginity, 55, 165
virtual reality, 105; also see cyberspace
vision, visual cortex, see cognitive function
volition, see cognitive function

Walton, John, 160, 167
weather, *xiv*, 40, 114, 118
Wellhausen, Julius, 194
witch of Endor, see Medium of Endor
wholism, see holism
widow of Nain, 146
widow of Zarephath, 24, 44, 79, 146
Wiles, Maurice, 120
woman, see human origins—creation of woman
Word of God, 194
    Book of God's word, Book of God's work, *xiv*, 113, 201
    two-edged sword, 49, 121
Wright, N.T., 24, 62, 88, 173, 175, 181–82, 187
writing of the Bible
    assembling, collating, 27–30
    canonization, 27–30
    cultural influence, 3, 7, 9–13, 20, 21
    during the exile, 7, 8–11
    re-writing during exile, 10–11, 157
    influence, Canaanite, 13

influence, Egyptian, 7, 13, 15–16, 32, 44, 126
influence, gender, 21, 22, 26
influence, Greek/Hellenism, 12, 14, 16, 18, 23–24, 32, 43, 44, 46, 48, 50–53, 55, 62, 126, 149, 157, 162–64
influence, Mesopotamian, 13, 21, 32, 44, 126, 149, 157, 159, 182
influence, NeoBabylonian, 13, 16, 44, 126, 149, 156–57, 159
influence, Zoroastrian, 11–12, 14, 24, 149182
inspiration, see divine inspiration
intertestamental period, 12, 155–56, 163–64
syncretism, 22
also see: Hebrew script; revelation; writing styles
writing styles, genres and tools
    allegory, 160, 167, 168
    analogy, 61, 74, 86, 88, 97, 100, 177, 191
    apologetics, 1, 189–90
    genealogy, 22, 166
    hagiography, 22
    metaphor, 41, 42, 44, 46, 65, 79, 90, 96, 124, 138, 147, 148, 156, 157, 162, 166–68, 177, 182, 186
    mythology, 14–20, 65, 158–60, 166; also see mythology
    parallelism, 82
    plagiarism, 20–22
    polemics, 21–22, 159
    propaganda, 22, 154, 171–72
    re-conceptualization, 161, 166
    simile, 162, 167–68

Xenophanes, 75

young earth, 70
Young, Davis A., 125

Zeus, 18, 19, 20, 38, 49, 172
Ziusudra, 19, 35, 36
Zoroastrian (Z—ism), 11–12, 14, 24, 133, 146, 149, 182, 184, 191

Scripture Index

Genesis

1	164, 167
1:26	15, 50
1:27	49, 50, 75
1:28	43
1:29	43
1:30	39, 124
2	49, 54, 167
2:8–14	158
2:21–22	39
2:15–22	50
2:7	16, 39, 77, 80
2:8	50, 164
2:9	159
2:12	160
2:13	159
2:17	121, 125, 182
2:19	159
3	18, 161, 187
3:22–24	43
3:3	77
3:8	159
3:19	16, 77, 135, 182
3:24	158, 159, 161
4:25–26	45
5:24	78, 139, 163
6:1–4	44
6:1	20
6:2	20
6:3	20, 45
6:17	39, 124
7:15	39, 124
7:22	39, 124
9:22–27	45
9:5	39
9:10	42
11:31	4
12	6
12:1	5
12:6	6
13:14–17	5
13:10	162
14	6
14:18	24
14:21	42
15:1	5
15:13	6
15:17	5
17:1–22	5
17:5	5
18	5, 6
18:1–8	6
18:27	17
20–21	6
20:13	5
23	6, 143
23:1–2	6
23:34–35	6
25:7–8	78, 142
25:9–10	143
25:8	39
25:17	39, 78, 142
25:18	160
26	6
31:19	5

Genesis (continued)

31:34	5
33	6
33:18–20	143
35	13
35:19–20	143
35:28–29	78, 142
35:2	5
35:18	39
35:29	39
37:34–35	142
37:35	139
39–50	6
42:38	142
44:29, 31	142
46	13
46:27	42
47:27	6
49:29–32	143
49:33	39, 142
50:12–14	143
50:24–26	143

Exodus

1:5	42
1:1–7	6
2:10	7
3:1—4:17	156
12	13
12:12	5
12:40	7
13:19	143
14:21	40
15:8	40
15:12	138
19	156
20	156
22:18	78, 134
28:12–19	160
28:15–30	160

Leviticus

17:11	39
17:14	39
19:31	78, 134
20:2–5	134
20:6	78, 134
20:27	78, 134

Numbers

6:5	40
11:17–25	43, 79
13:33	44, 45
16	141
16:23–33	138
16:30–34	140
16:32–34	138
16:30	137
16:33	137
20:24	78, 142
21:14	28
22:21–30	24
25:2	78, 134
26:9–11	141
26:58	141
27:18–20	43
27:13	78, 142
27:20	79
31:2	78, 142

Deuteronomy

1:28	42
6:5	82
8:25	140
12:23	39
18:10–11	78, 134
20:16	39
26:14	78, 134
28:25	140
32:50	78, 142

Joshua

5:1	42
6:26	134
7:5	42
10:13	28
10:16	42
10:40	39, 78
11:11	39, 78
14:28	42
24:2	4
24:32	143

Judges

8:32	78, 142
11:31	134
11:39	134
12:1–6	8
12:4–6	28, 195
13:5	40
16:17	40
17:6	9
20:1–48	8
21:25	9

1 Samuel

1:1	134
2:6	78, 137, 138
13:14	42
15:7	160
28:4–19	141
28:6–19	134
28:11–14	126
28:13	5, 44, 84
28:15	138

2 Samuel

3:33	140
4:12	134, 140
11:1	22
12:23	79, 137
14:14	79, 137
17:10	42
22:5	138
22:6	138

1 Kings

2:6	136
2:10	78, 142
11:41	28
11:33	5
11:43	78, 142
14:29	28
15:29	39, 78
16:34	134
17:17–22	146
17:21–22	44, 79
17:9	24

2 Kings

2:9–15	43, 79
2:23–24	141
2:11	79, 139
3:27	134
4:32–35	146
5:26	43, 79
5:1–19	24
8:24	78, 142
13:20–21	134, 146
19:12	157, 167
22:20	78
23:10	156
24:6	78, 142

1 Chronicles

1:1	166
6:22	141
20:1	22
29:28	78, 142
29:29	28

2 Chronicles

9:29	28
12:15	28
27:9	78, 142
29:12	157, 167
32:30	159
34:28	78, 142
36:22	9, 24

Ezra

1:1	9
1:1–8	24

Nehemiah

1:11	12

Job

2:4–6	80
3:11–17	174
3:13	138
7:7–10	140
7:3–15	142
7:9	137
7:10	137
10:8–9	16, 81
10:21	137, 138
11:8	138
14:7–12	81, 135, 145
14:10–12	137
14:1	135
14:12	138
14:22	43, 80
16:22	137
17:13–16	137, 138, 139
17:16	138
19:25	81, 136
19:26	145
24:19–20	138, 139
31:6	1, 132
37	114

37:9–10	40
38	114
38:17	138

Psalms

6:5	81, 135, 140, 145
6:5–6	137
8:5–8	80
8:4	1
8:5	5
9:15–17	142
9:14–15	138
9:17	139
13:3	138
16:3–4	134
16:9–10	81, 136, 145
16:10	138
18:4–16	80
18:4	138
18:16	137
30:3	80, 137, 138
30:9	81, 135, 145
31:17	137, 138, 139, 142
33:6	40
39:13	135
40:2	80, 137
40:8	40
41:5	135
42—49	141
48	156
49:14–15	138
49:15	80, 81, 136, 137, 145
49:17	137
51:10	43, 80
55:15	137, 138, 139
63:1	42
63:2	80
71:20	80, 137
73:23–26	81, 136, 145
84—89	141
86:13	80, 137, 138
88	137, 138, 140
88:10–14	140
88:3–4	141

88:11–12	81, 135, 145
88:4	138, 139
88:5	138, 140
88:6	138
88:7	138
88:8	138, 140
88:9	137
88:10	138
88:11	138
88:12	138
88:13	137, 138
89:48–49	138
89:48	138
90:3	135, 182
90:5	135
103:14	135, 182
104:29	39
106:28	134
107:10–22	80, 137
107:16–18	138
107:19–20	137
107:26	42
110:4	24
115:17	81, 135, 137, 138, 140, 145
137:9	22
138	83
139:14–15	16
139:13–16	81
139:8	138
139:13	39
146:4	80

Proverbs

5:5	137, 138
7:27	138, 175
12:10	42
15:24	138, 141, 142
20:27	40
22:18	40
23:14	141

Ecclesiastes

3:18–19	81, 135, 145
3:20	135, 182
7:17	140
9:5–6	135
9:7–10	138
9:10	138
12:7	44, 80, 81, 135, 136, 145
12:14	81, 145

Isaiah

1:14	42
5:14	138
7:11	138
7:14	29
8:19	79, 134
10:18	42, 79
14	187
14:9–10	79, 138
14:11	137, 138, 139
14:12	161
14:15	137, 138, 139
26:9	79, 83
26:14	79, 138, 140
26:19	80, 147 182
31:8–9	156
31:9	79
38:10	138, 139, 142
38:11	135
38:18	79, 135, 137, 138, 140
38:19	135
39:31–33	79
40:6	135
44:28	24
51:3	162
57:5–7	79, 134
57:9	79, 134, 137
59:10	79, 138, 156
66:24	79

Jeremiah

2:34	39
7:31	156
7:32	79, 156
7:33	140
8:1	140
16:4	140
16:6	140
19:5-7	79, 156
32:35	156
51:39	79, 138
51:57	79, 138
32:21-32	138
32:21-27	142
32:22-30	79, 138
32:21	138
32:23	138
36:26-27	80
36:26	43, 80
36:35	162
37:5-6	39
37	79, 147, 148
37:9	40
43:7	79, 134
43:9	79, 134

Lamentation

2:11	40
3:6	138

Ezekiel

8:3	43, 79, 80
9:10	40
11:19	43, 80
11:21	40
16:43	40
26:20	137
28	161, 162, 187
28:12	161
28:13	161
28:14, 16	158, 161
31	161, 187
31:14-17	138
31:15-17	139, 142
31:8-9	158
31:10-12	162
31:3, 5	161
31:6	162
31:8	161, 162
31:15	138
31:16	138, 158
31:17	162
31:18	158
32:18-32	139
32:19-32	139

Daniel

1:1-6	12
1:19	12
8:2	43, 80
12	146
12:1-3	25, 81, 84, 146, 147, 165
12:2	174
12:3	153
12:13	81, 84, 146, 147

Joel

2:3	162

Amos

1:5	157, 167
9:2	138

Jonah

2:1-9	137
2:3-8	137, 138
2:2	137
2:5	79
2:6	137, 138

Habakkuk

2:5	138

Zechariah

12:1	39, 79

Matthew

1:23	29
2:1–2	24
4:19	172
5:38–48	193
5:29–30	156
5:22	156
6:10	172
10:5–6	192
10:28	82, 156
11:12	168, 172
12:40	150
13:44–45	168
13:47, 52	168, 172
13:31	167, 172
13:33	167, 172
15:24	192
16:21	150
17:1–13	151
17:22	150
22:2	172
22:30	176, 177
22:32	174
22:37	82
23:15, 33	156
24:15, 30, 36	176
26:14	82
26:29	176, 181
26:61	150
27:50–53	146
27:50	82
27:63	150
28:8–9	150
28:9	176
28:19	192

Mark

1:15	168
8:31	150
9:43, 45	156
9:3	151
12:26–27	174
12:25	176, 177
12:30	82
14:58	150
15:37	39, 82
16:12	150
16:15	170
16:19	150

Luke

1:46–47	48, 83
2:14	171
4:26	24
4:27	24
7:11–15	146
8:49–55	146
8:41–42	146
9:28–36	151
9:22	150
10	187
10:18	161
10:27	82
12:5	156
16:19–31	174
16:22	173
17:20–21	168
20:35	176, 177
20:38	174
23:40–43	174
23:43	150, 158, 163, 175
23:46	39, 82
24:15–31	176
24:37–40	150
24:13–16	151
24:41–43	176
24:45–47	29
24:29–30	176
24:36	151
24:39	176
24:51	151, 176

John

2:19	150
3:3	121, 125
3:4	121
3:9	121
11:1–44	146
11:11–14	174
14:1–3	150
14:2–3	181
19:30	82
20:14–18	176
20:14	176
20:17	150, 176, 181
20:19	176
20:20	150, 151
20:22	83, 125
20:26	176
20:27	150

Acts

1:9–10	151
1:9	176
2:5, 14, 46	192
5:5	82
5:10	82
6:1–7	192
7:15–16	144
7:6	7
7:21	7
7:22	7
7:60	174
8:26–39	24
9:36–41	146
12:23	82
13:20	7
15	192
15:29	13, 192
17:22–30	24
17:2	192
20:9–10	146
21:13	153
22:6–10	153
23:8	150
26:22–23	153

Romans

1:20	190
5:6	190
8:26–27	122
8:29	75, 153

1 Corinthians

3:16	83
5:1–4	82
5:3	82
5:5	82
8:4–13	13, 192
8:6	123
9:1	153
11:1–16	23
11:14–15	13, 40, 193
11:7	75
14:14	103
15:12–58	177
15:17–19, 32	152
15:42–44	177
15:52–54	153
15:8	153
15:12	177
15:18	174
15:20	153, 174
15:29	175
15:42	153
15:44	153, 181
15:49	75, 153
15:50	153
15:51	174
15:53	153

2 Corinthians

3:18	75
4:16	82
5:1–10	174
5:4	153
5:8	82, 186
8:9	123
12:2–4	163, 164

Galatians

3:17	7
3:27–28	177, 178
4:4	190

Ephesians

4:22–24	122
4:8	175
4:24	75

Philippians

1:21–24	174
1:8	40
2:6–7	123
2:17	153
3:20–21	153

Colossians

1:15–17	123
1:15, 18	153

1 Thessalonians

4:3—5:11	177
4:15	174
4:16	153
4:16	153
5:10	174
5:23	49, 82, 83

1 Timothy

6:7	123

2 Timothy

4:6–8	153

Hebrews

1:3	75
2:10	153
4:12	49, 83, 121
5:6	24
5:10	24
6:20	24
7:1–2	24
7:10–11	24
11:5	163
11:24	7
11:35	153
12:14	174
13:8	195

James

1:18	153
2:26	82
5:5	156

1 Peter

3:18–19	83, 174, 177
3:19–20	175

Revelation

2:7	157, 163, 164
6:9–11	174
11:15	173
18:13	83
21:15–18	181
21:27	174
22	163
22:1–5	158
22:1–4	157, 163
22:18–19	27

Ancient Document Index

Ancient Near Eastern Documents

Atrahasis, 15, 19, 35, 36
Enuma Elish, 4, 15
Hymn to E'engura, 15
Enki and Ninma, 15
KAR4, 15
Hymn of Akhenatan, 37
Song of the Hoe, 15
Urdeuteronomium, 10
Annals of Samuel the Seer, 28
the Records of Nathan the Prophet, 28
the Records of Gad the Seer, 28
the Visions of Iddo the Seer, 28
the Records of Shemaiah the Prophet, 28
the Book of the Wars of the Lord, 28
the Book of Jashar, 28
the Annals of the Acts of Solomon, 28
the Annals of the Kings of Judah, 28

Apocrypha

Books of Maccabees, 148, 153, 154, 176
Tobit, 29, 48, 81
Ecclesiasticus, 29
Baruch, 164
1 and 2 Esdras, 29

Pseudepigrapha

Book of Ezra, 164
Books of Enoch, 48, 81, 148, 155, 157, 160, 161, 164
Book of Jubilees, 48, 81, 163
Gospel of Peter, 29
Pseudo-Ezekiel, 148, 154, 161
Sibylline Oracles, 48, 81
Testament of Abraham, 163–64
Testament of Dan, 164
Testament of Job, 80

Dead Sea Scrolls

Messianic Apocalypse, 148

Rabbinic Writings

Baruch, 164

Greco-Roman Writings

Jewish Antiquities, 164
Theogony, 15

Early Christian Writings

Apocalypse of Peter, 29
Biblical Antiquities of Pseudo-Philo, 164
3rd Corinthians, 30
Epistle to the Laodiceans, 29
Epistle of Barnabas, 29
Gospel of the Ebionites, 29
Gospel of Judas, 29
Gospel of Philip, 29
Gospel of the Nazarenes, 29
Gospel of Thomas, 29
Gospel of Truth, 29
Letter to the Laodiceans, 30
Revelation of Paul, 29
Septuagint, 29
Shepherd of Hermas, 29

www.ingramcontent.com/pod-product-compliance
Lightning Source LLC
Chambersburg PA
CBHW050848230426

43667CB00012B/2205